MECHANICAL & ELECTRICAL SYSTEMS

Marc Schiler & Consulting Editor Shakeel Ahmed

KAPLAN AEC EDUCATION

President: Roy Lipner
Vice President of Product Development and Publishing: Evan M. Butterfield
Editorial Project Manager: Jason Mitchell
Director of Production: Daniel Frey
Production Editor: Caitlin Ostrow
Creative Director: Lucy Jenkins

Published by Kaplan AEC Education
30 South Wacker Drive, Suite 2500
Chicago, IL 60606-7481
(312) 836-4400
www.kaplanaecarchitecture.com

Printed in the United States of America.

07 08 09   10 9 8 7 6 5 4 3 2 1

CONTENTS

WELCOME

Thank you for choosing Kaplan AEC Education for your ARE study needs. We offer updates every January to keep abreast of code and exam changes and to address any errors discovered since the previous update was published. We wish you the best of luck in your pursuit of licensure.

ARE OVERVIEW

Since the State of Illinois first pioneered the practice of licensing architects in 1897, architectural licensing has been increasingly adopted as a means to protect the public health, safety, and welfare. Today, all U.S. states and Canadian provinces require licensing for individuals practicing architecture. Licensing requirements vary by jurisdiction; however, the minimum requirements are uniform and in all cases include passing the Architect Registration Exam (ARE). This makes the ARE a required rite of passage for all those entering the profession, and you should be congratulated on undertaking this challenging endeavor.

Developed by the National Council of Architectural Registration Boards (NCARB), the ARE is the only exam by which architecture candidates can become registered in the United States or Canada. The ARE assesses candidates' knowledge, skills, and abilities in nine different areas of professional practice, including a candidate's competency in decision making and knowledge of various areas of the profession. The exam also tests competence in fulfilling an architect's responsibilities and in coordinating the activities of others while working with a team of design and construction specialists. In all jurisdictions, candidates must pass the nine divisions of the exam to become registered.

The ARE is designed and prepared by architects, making it a practice-based exam. It is generally not a test of academic knowledge, but rather a means to test decision-making ability as it relates to the responsibilities of the architectural profession. For example, the exam does not expect candidates to memorize specific details of the building code, but requires them to understand a model code's general requirements, scope, and purpose, and to know the architect's responsibilities related to that code. As such, there is no substitute for a well-rounded internship to help prepare for the ARE.

Exam Format

The ARE consists of nine divisions: three graphic and six multiple-choice. The multiple-choice divisions are timed and contain differing numbers of questions. The number of questions and time limit for each division is outlined in the table below. For detailed information on the graphic divisions, refer to the study guides for those divisions.

Division	Questions	Hours
Pre-Design	105	2.5
General Structures	85	2.5
Lateral Forces	75	2
Mechanical and Electrical Systems	105	2
Building Design/ Materials & Methods	105	2
Construction Documents & Services	115	3

The exam presents questions individually. Candidates may answer questions, skip questions, or mark questions for further review. Candidates may also move backward or forward within the exam using simple on-screen icons. The technique of marking questions for further review is a useful tool.

ARCHITECTURAL HISTORY

Questions pertaining to the history of architecture appear in all of the multiple-choice divisions. The prominence of historical questions will vary not only by division but also within different versions of the exam for each division. In general, however, history tends to be lightly tested, with approximately three to seven history questions per division, depending upon the total number of questions within the division. One aspect common to all the divisions is that whatever history questions are presented will be related to that division's subject matter. For example, a question regarding Chicago's John Hancock Center and the purpose of its unique exterior cross bracing may appear on the Lateral Forces exam.

Though it is difficult to predict how essential your knowledge of architectural history will be to passing any of the multiple-choice divisions, it is recommended that you refer to a primer in this field—such as Kaplan's *Architectural History*—before taking each exam, and that you keep an eye out for topics relevant to the division for which you are studying. It is always better to be overprepared than taken by surprise at the testing center.

Actual appointment times for taking the exam are slightly longer than the actual exam time, allowing candidates to check in and out of the testing center. All ARE candidates are encouraged to review NCARB's *ARE Guidelines* for further detail about the exam format. These guidelines are available via free download at NCARB's Web site (*www.ncarb.org*).

Question Format

It is important for exam candidates to familiarize themselves not only with exam content, but also with question format. Familiarity with the basic question types found in the ARE will reduce confusion, save time, and help you pass the exam. The multiple-choice divisions contain three basic question types.

The first and most common type is a straight-forward multiple-choice question followed by four choices (A, B, C, and D). Candidates are expected to select the correct answer. This type of question is shown in the following example.

Which of the following cities is the capital of the United States?

A. New York

B. Washington, DC

C. Chicago

D. Los Angeles

The second type of question is a negatively worded question. In questions such as this, the negative wording is usually highlighted using all caps, as shown below.

Which of the following cities is NOT located on the west coast of the United States?

A. Los Angeles

B. San Diego

C. San Francisco

D. New York

The third type of question is a combination question. In a combination question, more than one choice may be correct; candidates must select from combinations of potentially correct choices. An example of a combination question is shown on the next page.

NEW TO THE EXAM

ARE 3.1

In November 2005 NCARB released *ARE Guidelines* Version 3.1, which outlines changes to the exam that are effective as of February 2006. These new guidelines primarily detail changes for the Site Planning division, which now combines the site design and site parking vignettes as well as the site zoning and site analysis vignettes. For more details about these changes, please refer to the study guides for the graphic divisions.

The new guidelines mean less to those preparing for multiple-choice divisions. Noteworthy points are outlined below.

- All division statements and content area descriptions are unchanged for the multiple-choice divisions.

- The number of questions and time limits for all exams are unchanged.

- The list of codes and standards candidates should familiarize themselves with has been reduced to those of the International Code Council (ICC), the National Fire Protection Association (NFPA), and the National Research Council of Canada.

- A statics title has been removed from the reference list for General Structures.

Rolling Clock

A rolling clock went into effect January 1, 2006. Candidates must now pass all nine ARE divisions within a five-year period. Additionally, NCARB has instituted a set of "transitional rules" for candidates already in the process of taking the ARE when the clock went into effect. See the new guidelines or visit the NCARB Web site for more detailed information.

Which of the following cities are located within the United States?

    I. New York

    II. Toronto

    III. Montreal

    IV. Los Angeles

A. I only

B. I and II

C. II and III

D. I and IV

Recommendations on Exam Division Order

NCARB allows candidates to choose the order in which they take the exams, and the choice is an important one. While only you know what works best for you, the following are some general considerations that many have found to be beneficial:

1. The Building Design/Materials & Methods and Pre-Design divisions are perhaps the broadest of all the divisions. Although this can make them among the most intimidating, taking these divisions early in the process will give a candidate a broad base of knowledge and may prove helpful in preparing for

subsequent divisions. An alternative to this approach is to take these two divisions last, since you will already be familiar with much of their content. This latter approach likely is most beneficial when you take the exam divisions in fairly rapid succession so that details learned while studying for earlier divisions will still be fresh in your mind.

2. The Construction Documents & Services exam covers a broad range of subjects, dealing primarily with the architect's role and responsibilities within the building design and construction team. Because these subjects serve as one of the core foundations of the ARE, it may be advisable to take this division early in the process, as knowledge gained preparing for this exam can help in subsequent divisions.

3. The General Structures and Lateral Forces divisions cover related and overlapping subjects. Take them consecutively, and take General Structures first, since it is broader and addresses fundamental principles necessary for success in Lateral Forces.

4. The three graphic divisions all use an identical software platform and employ similar graphic drawing tools. Because becoming fluent with this software is crucial to passing these exams, take the three graphic divisions sequentially.

5. The Mechanical & Electrical Systems and Building Technology exams cover loosely related material. As such, it is often beneficial to take these two exams consecutively.

6. Take exams that particularly concern you early in the process. NCARB rules prohibit retaking an exam for six months. Therefore, failing an exam early in the process will allow the candidate to use the waiting period to prepare for and take other exams.

EXAM PREPARATION

Overview

There is little argument that preparation is key to passing the ARE. With this in mind, Kaplan has developed a complete learning system for each exam division, including study guides, question-and-answer handbooks, mock exams, and flash cards. The study guides offer a condensed course of study and will best prepare you for the exam when utilized along with the other tools in the learning system. The system is designed to provide you with the general background necessary to pass the exam and to provide an indication of specific content areas that demand additional attention.

In addition to the Kaplan learning system, materials from industry-standard documents may prove useful for the various divisions. Several of these sources are noted in the "Supplementary Study Materials" section below.

Understanding the Field

The subject of mechanical and electrical systems rarely falls directly under the responsibility of the architect, but such systems are largely responsible for achieving the most basic objectives in architecture—health, safety, comfort, and convenience for occupants. As such, the subject of mechanical and electrical systems must be understood in order for architects to properly integrate these systems within their designs and to permit the architect to interact constructively with other members of the building design team.

Understanding the Exam

The mechanical and electrical systems exam covers subjects including sustainability, plumbing systems, thermal protection systems, lighting, acoustics, mechanical, fire protection, and transportation systems. A basic understanding of the function and purpose of these systems is needed in order to succeed with this exam. However, the exam does not extensively test a candidate's knowledge of the detailed design of such systems, rather their components and how they function. Environmental and energy conservation is a major focus on the exam material. Further, candidates should possess knowledge of how plumbing affects health factors as well as life safety systems.

Preparation Basics

The first step in preparation should be a review of the exam specifications and reference materials published by NCARB. These statements are available for each of the nine ARE divisions to serve as a guide for preparing for the exam. Download these statements and familiarize yourself with their content. This will help you focus your attention on the subjects on which the exam focuses.

Though no two people will have exactly the same ARE experience, the following are recommended best practices to adopt in your studies and should serve as a guide.

Set aside scheduled study time.
Establish a routine and adopt study strategies that reflect your strengths and mirror your approach in other successful academic pursuits. Most importantly, set aside a definite amount of study time each week, just as if you were taking a lecture course, and carefully read all of the material.

Take—and retake—quizzes.
After studying each lesson in the study guide, take the quiz found at its conclusion. The quiz questions are intended to be straightforward and objective. Answers and explanations can be

found at the back of the book. If you answer a question incorrectly, see if you can determine why the correct answer is correct before reading the explanation. Retake the quiz until you answer every question correctly and understand why the correct answers are correct.

Identify areas for improvement.
The quizzes allow you the opportunity to pinpoint areas where you need improvement. Reread and take note of the sections that cover these areas and seek additional information from other sources. Use the question-and-answer handbook and CD-ROM test bank as a final tune-up for the exam.

Take the final exam.
A final exam designed to simulate the ARE follows the last lesson of each study guide. Answers and explanations can be found on the pages following the exam. As with the lesson quizzes, retake the final exam until you answer every question correctly and understand why the correct answers are correct.

Use the flash cards.
If you've purchased the flash cards, go through them once and set aside any terms you know at first glance. Take the rest to work, reviewing them on the train, over lunch, or before bed. Remove cards as you become familiar with their terms until you know all the terms. Review all the cards a final time before taking the exam.

Supplementary Study Materials

In addition to the Kaplan learning system, materials from industry-standard sources may prove useful in your studies. Candidates should consult the list of exam references in the NCARB guidelines for the council's recommendations and pay particular attention to the following publications, which are essential to successfully completing this exam:

- International Code Council (ICC)
  International Building Code
- National Fire Protection Association
  Life Safety Code (NFPA 101)

Test-Taking Advice

Preparation for the exam should include a review of successful test-taking procedures—especially for those who have been out of the classroom for some time. Following is advice to aid in your success.

Pace yourself.

Each division allows candidates at least one minute per question. You should be able to comfortably read and reread each question and fully understand what is being asked before answering.

Read carefully.

Begin each question by reading it carefully and fully reviewing the choices, eliminating those that are obviously incorrect. Interpret language literally, and keep an eye out for negatively worded questions.

Guess.

All unanswered questions are considered incorrect, so answer every question. If you are unsure of the correct answer, select your best guess and/or mark the question for later review. If you continue to be unsure of the answer after returning the question a second time, it is usually best to stick with your first guess.

Review.

The exam allows candidates to review and change answers within the time limit. Utilize this feature to mark troubling questions for review upon completing the rest of the exam.

Reference material.

The General Structures and the Mechanical & Electrical Systems divisions include reference materials accessible through an on-screen icon. These materials include formulas and other reference content that may prove helpful when answering questions in these divisions. Note that candidates may *not* bring reference material with them to the testing center.

Calculator.

Candidates must bring their own calculator to the testing center. Note that only nonprogrammable, noncommunicating, nonprinting calculators are allowed. Candidates will need only a basic scientific calculator with trigonometry functions.

Best answer questions.

Many candidates fall victim to questions seeking the "best" answer. In these cases, it may appear at first glance as though several choices are correct. Remember the importance of reviewing the question carefully and interpreting the language literally. Consider the example below.

> Which of these cities is located on the east coast of the United States?
>
> A. Boston
> B. Philadelphia
> C. Washington, DC
> D. Atlanta

At first glance, it may appear that all of the cities could be correct answers. However, if you interpret the question literally, you'll identify the critical phrase as "on the east coast." Although each of the cities listed is arguably an "eastern" city, only Boston sits on the Atlantic coast. All the other choices are located in the eastern part of the country, but are not coastal cities.

ACKNOWLEDGMENTS

This course was written by Marc Schiler. Mr. Schiler, who has a bachelor's degree from USC and a master's degree from Cornell, has written, lectured, and consulted extensively on environmental controls and energy in buildings. He currently is a professor at USC.

Material on the subject of sustainable design was written by Jonathan Boyer, AIA. Mr. Boyer is a principal of the firm Boyer Associates Ltd. in Chicago. He is a graduate of the University of Pennsylvania (BA) and Yale University (MArch). His firm has focused on sustainable design and environmental planning for more than 30 years with projects throughout the United States.

This introduction was written by John F. Hardt, AIA. Mr. Hardt is a principal of the firm Andrews Architects, Inc. in Columbus, Ohio. He is a graduate of Ohio State University (MArch) and has been in practice for more than 12 years.

ABOUT KAPLAN

Thank you for choosing Kaplan AEC Education as your source for ARE preparation materials. Whether helping future professors prepare for the GRE or providing tomorrow's doctors the tools they need to pass the MCAT, Kaplan possesses more than 50 years of experience as a global leader in exam prep and educational publishing. It is that experience and history that Kaplan brings to the world of architectural education, pairing unparalleled resources with acknowledged experts in ARE content areas to bring you the very best in licensure study materials.

Only Kaplan AEC offers a complete catalog of individual products and integrated learning systems to help you pass all nine divisions of the ARE. Kaplan's ARE materials include study guides, mock exams, question-and-answer handbooks, video workshops, and flash cards. Products may be purchased individually or in division-specific learning systems to suit your needs. These systems are designed to help you better focus on essential information for each division, provide flexibility in how you study, and save you money.

To order, please visit *www.KaplanAEC.com* or call (800) 420-1429.

ACKNOWLEDGMENTS

Kaplan AEC is grateful for the cooperation of many experts, professionals, organizations, and associations in the creation of this book. The following images are reproduced with the permission of the copyright holders.

Title	*Page*
Table 1.1—Specific Heat (C_p)	44
Table 1.2—Air Space Resistances	47
Table 1.3—Thermal Resistances	49
Table 1.6—Btuh for Various Activities	56
Table 1.7—CLTD Wall Groups	58
Table 1.8—CLTD Values	59
Table 1.9—Shading Coefficients	60
Table 1.10—Solar Intensity (S_g) for 40 North Latitude	60
Comfort Zone	69
Table 3.1—Friction Loss	105

Copyright 1989, American Society for Heating, Refrigerating and Air-Conditioning Engineers, Inc. (www.ashrae.org). Reprinted by permission from 1989 ASHRAE Handbook of Fundamentals. This text may not be copied nor distributed in either paper or digital form without ASHRAE's permission.

Psychometric Chart	67
Cold Climate Plot	75

Courtesy of Carrier Corporation

Veiling Reflections	132

Used by permission of Pilkington North American, Inc.

Table 6.5—Reverberation Times	153

Reprinted courtesy of BBN Technologies.

Table 7.2—Fixture Units Per Fixture Type	39
Table 7.3—Pipe Sizes for Fixture Unit Totals	39

Portions of this document reproduce tables from the *2003 International Plumbing Code,* International Code Council, Falls Church, Virginia. Reproduced with permission. All rights reserved.

INTRODUCTION

The ARE requires students to incorporate building codes and other regulatory requirements into system design. Below you will find an outline of the most relevant portions of the *Uniform Building Code* and *Uniform Plumbing Code*. Please note that your state or municipality may have amended these codes, or adopted other codes (such as the *International Building Code*). Check with your state or municipality for further information.

CODES AND REGULATIONS

1.0 Uniform Building Code/Local Code Amendments.

   a. Group B, division 2 occupancy (offices, retail, etc.)

     1) Minimum mechanical ventilation required for interior spaces and other areas void of means of natural ventilation—15 cfm minimum of outside air (OSA) per person (for Groups A, B, E, F, H, I, M & S). *Note:* ASHRAE Standard 62 stipulates 20 cfm per person.

     2) Toilet exhaust—Fully openable exterior window or vertical duct to outside, or mechanical exhaust minimum of four air changes/hour. Discharge to outside shall be at least three feet from any openings that allow air entry into occupied portion of the building.

     3) Storage rooms for Class I, II, or III-a liquids-exhaust six air changes at or near floor level.

     4) Parking garages (Group S occupancy)—mechanical exhaust minimum of 1.5 cfm/square feet of gross area when openness requirements cannot be met. *Alternate:* Exhaust 14,000 cfm per operating vehicle based on not less than 2.5 percent of garage capacity. CO sensing system may cycle fans to maintain concentrations between 50 ppm (eight hour maximum average) & 200 ppm (one hour maximum).

     5) Rooms containing fuel fired heating equipment where gas input of larges piece of equipment is in excess of 400,00 BTU per hour shall be separated with not less than a one-hour occupancy separation. Two exits required if room is larger than 500 sq. ft. (For Group A occupancies 0 exterior opening from such rooms shall be a minimum of 10' from other doors and windows openings of some building).

   b. Group R occupancy

     1) Guest rooms and habitable rooms within a dwelling unit: If mechanically ventilated—minimum of two air changes per hour with minimum 15 CFM per person outside air in occupied areas (including guest rooms).

     2) Toilets, bathrooms, and similar rooms—natural ventilation by openings to outside, or mechanical: Exhaust—minimum of five air changes per hour. Exhaust discharge shall be a minimum of three feet from any building openings.

   c. Atriums

     1) Mechanical smoke removal system.

   d. "High-rise" buildings over 75 feet from grade, Type I construction.

     1) Smoke control required.

     2) Elevator hoistways:

       i) Do not vent through elevator machine room. Vent top of shaft immediately below the elevator machine room with opening of 3-

1/2 percent of area of shaft but no less than three square feet per elevator.

3) Stairways and vestibules:

i) Mechanically pressurize stairwell to 0.05" W.G. with reference to vestibule.

ii) Exhaust 2,500 CFM with barometric damper at top of stairwell.

iii) Vestibule pressurized to 0.05" SP with reference to floor.

e. Chimneys, fireplaces, and barbecues

1) Terminations:

i) Residential—two feet above roof and two feet above any part of the building within ten feet.

ii) Building heating and industrial low-heat—three feet above roof and two feet above any part of the building within ten feet.

iii) Watch out for grouping chimney termination if any downdraft can occur.

1.1 Uniform mechanical code/local code amendments.

a. Equipment—general

1) Central heating furnaces not listed for alcove or closet installation shall be installed in a room or space having a volume at least 12 times that of the furnace.

2) Central heating boilers shall have a room volume at least 16 times that of the boiler. Room volume shall assume a maximum 8' high ceiling for purposes of the calculation.

Example:

If a hot water boiler were 4' wide x 7' long x 5'6" high, room size would be

$4 \times 7 \times 5.5 \times 16 = 2,464$ cubic ft./8' high = 308 sq. ft., or

10' x 30', or 15' x 21'. *Very large.*

Note: two exists are required for boiler rooms larger than 500 sq. ft. with largest piece of fuel-fired equipment in excess of 400,000 BTU per hour input.

3) Access—Provide 30 inches minimum of working space and platform for inspection, service, repair, and replacement of appliances.

b. Commercial hoods and kitchen ventilation.

1) Type I hood—kitchen hood—grease and smoke.

2) Type II hood—general hood—steam, vapor, heat, and odors.

3) Ducts: Type I hood—16 ga. Continuously welded ductwork.

i) Minimum ¼" /foot slope toward hood.

ii) Fire-rated enclosure to outside (two-hours enclosure for Type I and II buildings)

iii) Air velocity 1,500–2,500 FPM

iv) Grease exhaust ducts shall extend at least two feet above the roof, discharge 40" minimum above roof, at least ten feet from parts of the same building, property line or air intake opening and ten feet above the adjoining grade level.

c. Duct system

1) Fire-rated corridors shall not be used to convey air to or from rooms.

2) Gas venting systems (flues) shall not extend into or through ducts or plenums.

3) Automatic shut-off: each heating or cooling system providing air in excess of 2,000 CFM shall be equipped with

an auto shut-off activated by a smoke detector. Smoke detectors shall be installed in the main supply duct.

4) Product conveying ducts.

  i) Definition: ducting used to convey refuse, dust, fumes, smoke, mists, vapors (flammable or corrosive liquids), and noxious or toxic gases.

  ii) Termination:

    (i) Explosive or flammable vapors: 30 feet from property line and building openings.
10 feet from openings into building.
6 feet from exterior walls or roof.
10 feet above adjoining grade.

    (ii) Corrosive, noxious, or toxic vapors:
10 feet from property line.
3 feet from exterior walls or roof.
10 feet from openings into building.
10 feet above adjoining grade.

    (iii) Environmental air ducts:
Domestic kitchen range exhaust: Must terminate outside of building.
Must have backdraft dampers.
Termination: three feet from property line and openings into building.

d. Combustion air:

1) General: fuel burning equipment shall be assured a sufficient supply of combustion air.

2) Typical requirement: two openings, each 1 sq. in./4,000 btu/hr shall be located within the upper 12" of the enclosure and the lower 12" of the enclosure. See UMC table for specific applications.

e. Venting of appliances:

1) Type B vents for gas appliances with draft hoods only.

2) Termination – type B & L vent: minimum of two feet above roof and no less than ten feet from any portion of a building which extends at an angle of more than 45 degrees from the horizontal.

3) Vent terminals: vents shall terminate no less than four feet below or four feet horizontally nor less than one foot above any door, window, or air inlet.

4) Vent Termination from property line: no less than three feet above any outside or make-up air inlet located within ten feet nor less than four feet from any property line.

5) Length-pitch: gravity vents shall extend generally in the vertical direction with offsets not exceeding 45°. Horizontal runs (angles greater than 45°) shall not exceed 75 percent of the vertical height.

6) Vents for building heating and industrial-type low-heat equipment shall be three feet above roof opening and two feet above any part of the building within ten feet.

f. Cooling systems

1) Outside air intakes—prohibited sources:

  i) Not closer than ten feet from any appliance or plumbing vent outlet unless vent is three feet above outside air inlet.

  ii) Ten feet minimum from HVAC system exhaust.

  iii) Where it will pick up objectionable odors, fumes.

iv) A hazardous or unsanitary location.

v) A refrigeration machinery room.

vi) A closet.

vii) Less than ten feet above abutting public way or driveway.

2) Return air—shall not be discharged from one dwelling unit to another through the cooling system.

3) Unobstructed access and passageway not less than 2'-6" width by 6'-6" high shall be provided for A/C units and fan equipment on the control or servicing side.

4) When a cooling unit is located above a ceiling or in an attic space, a second condensate pan shall be installed with its drain piped to a point that can be easily observed.

g. Mechanical refrigerating equipment

1) Ventilation of equipment rooms for condensing units (other than machinery rooms):

i. Gravity ventilation openings of not less than two square feet net.

ii. Mechanical exhaust system with three air changes.

2) Exits: Refrigerant compressors of more than five HP shall be located at least ten feet from an exit unless separated by a one-hour occupancy separation.

3) Refrigeration machinery room

i.) One-hour construction.

ii) Required for cooling units in excess of 100 HP.

iii) Ventilation by mechanical means based on HP of refrigeration equipment or gravity ventilation openings with ½ of area within 12 inches of the floor and ½ within 12 inches of the ceiling.

iv) Unobstructed working space of 3'-0" x 6'-8" high shall be provided around two adjacent sides of all moving machinery.

v) Mechanical ventilation shall provide 12 air changes/hr. (ACH) exhausting a minimum of 20 feet from any property line, exterior door, window, or opening. An emergency refrigeration control switch shall be provided to shut off electrically operated refrigeration machinery. Locate switch within ten feet of room exit.

h. Gravity ventilation of machinery rooms shall provide a minimum of 17 sq. ft. of opening to the outside per 100 HP.

1.3 Uniform Plumbing Code

a. Galvanized wrought iron or galvanized steel pipe is not allowed to be used underground.

b. PVC or ABS pipes are allowed in underground for structures limited to three floors above grade.

c. All plumbing fixtures shall be drained by gravity. Drain pipes shall be sloped minimum 1/8" per foot in the direction of flow for pipes four inch diameter or larger. Other pipes shall be sloped at ¼" per foot.

d. Clean outs shall be provided for in sewer pipes inside the building at 100 feet intervals and each change of direction.

e. Condenser drain for HVAC equipment shall be connected to water pipe indirectly with air gap.

f. Natural gas piping exposed to weather shall be galvanized.

1.4 Electrical Code (NEC)

a. Water pipes not directly servicing the electrical room HVAC equipment cannot be installed in the electrical room.

b. Ductwork (except supply/exhaust ducts serving the electrical room) should not be installed in the electrical room.

c. Provide at least 36 inches in front of electrical panels and switchboards for access and maintenance.

d. Provide emergency lighting to maintain 0.5 lightcandles in commercial buildings.

e. Computer rooms, server rooms, and network equipment rooms are required to have one-hour fire rated construction, and generally require an emergency power switch near the room entrance.

f. New buildings may require new electrical transformers from local power suppliers.

g. Transformer rooms, when installed in a building, require three-hour fire rated construction. These rooms require direct access to the street with a 10-foot door. Transformer rooms may require independent ventilation.

h. For commercial buildings, provide spaces for electrical and telephone rooms. Electrical rooms should be located on each floor to avoid longer conduit runs and minimize power loss.

SUSTAINABLE DESIGN

by Jonathan Boyer, AIA

This section is meant to prepare you for the *"green" architecture, sustainability,* and *new material technologies* topics that NCARB has introduced to several of the multiple-choice divisions of the ARE 3.0.

INTRODUCTION

Architects can no longer assume that buildings function independently of the environment in which they are placed. In the late 1800s the machine age offered the lure of buildings that were self-sufficient and independent of their natural surroundings—"The Machine for Living," as LeCorbusier proclaimed.

In the middle of the 20th century, the promise of endless and inexpensive nuclear energy lured architects into temporarily ignoring the reality of the natural elements affecting architectural design. Why worry about natural systems if energy was going to be infinite and inexpensive? Glass houses proliferated.

Energy is not free, the global climate is changing, and the viability of natural ecosystems is diminishing. Architects are designing structures that affect all these natural ecosystems. Much as Marcus Vitruvius wrote thousands of years ago that architects must be sensitive to the local environment, architects are returning to study the virtues of tuning to natural systems. Contemporary architects must combine their knowledge of the benefits of natural systems with the understanding of the selective virtues of contemporary innovative technologies.

This lesson focuses on the fundamental principles of environmental design that have evolved over the thousands of years that humans have been creating spatial solutions.

HISTORY OF SUSTAINABLE DESIGN

In early human history, builders of human habitats used materials that occurred naturally in the earth, such as stone, wood, mud, adobe bricks, and grasses. With nomadic tribes and early civilizations, the built environment made little impact on the balance of natural elements. When abandoned, the grass roof, adobe brick, or timber beam would slowly disintegrate and return to the natural ecosystem. Small human populations and the use of natural materials had very little impact on a balanced natural ecosystem.

But as human populations expanded and settlements moved into more demanding climates,

natural materials were altered to become more durable and less natural. In fact, archeological finds demonstrate some of the human creations that are not easily recycled into the earth; fired clay, smelted ore for jewelry, and tools are examples of designs that will not easily reintegrate into the natural ecosystem. These materials may be reprocessed (by grinding, melting, or reworking) into other human creations, but they will never be natural materials again.

As human populations expanded, there is strong evidence that some civilizations outgrew their natural ecosystem. When overused, land became less fertile and less able to support crops, timber, and domesticated animals necessary for human life. The ancient solution was to move to a more desirable location and use new natural resources in the new location, abandoning the ecologically ruined home site.

The realization that global natural resources are limited is an age-old concept. The term *conservation,* which came into existence in the late 19th century, referred to the economic management of natural resources such as fish, timber, topsoil, minerals, and game. In the United States, at the beginning of the 20th century, President Theodore Roosevelt and his chief forester, Gifford Pinchot, introduced the concept of conservation as a philosophy of natural resource management. The impetus of this movement created several pieces of natural legislation to promote conservation and increased appreciation of America's natural resources and monuments.

In the middle of the 1960s, Rachel Carson published *Silent Spring,* a literary alarm that revealed the reality of an emerging ecological disaster—the gross misunderstanding of the value and hazards of pesticides. The pesticide DDT and its impact on the entire natural ecosystem was dramatic; clearly, some human inventions were destructive and could spread harm throughout the ecosystem with alarming speed and virulence. Birds in North America died from DDT used to control malaria in Africa. Human creations were influenced by the necessities of the natural cycles of the ecosystem. Human toxic efforts could no longer be absorbed by the cycles of nature. Human activities became so pervasive and potentially intrusive that there needed to be a higher level of worldwide ecological understanding of the risk of disrupting the ecosystem.

Architects, as designers of the built environment, realize the ecological impact of their choices of architectural components, such as site selection, landscaping, infrastructure, building materials, and mechanical systems. The philosophy of sustainable design encourages a new, more environmentally sensitive approach to architectural design and construction.

There are many credos for the approach to a new sustainable design. Some architectural historians maintain that the best architects (Vitruvius, Ruskin, Wright, Alexander) have always discussed design in terms of empathy with nature and the natural systems. Now it is evident that all architects should include the principles of sustainable design as part of their palette of architectural best practices.

PRINCIPLES OF SUSTAINABLE DESIGN

The tenets outlined below indicate why it is necessary to maintain the delicate balance of natural ecosystems.

1. In the earth's ecosystem (the area of the earth's crust and atmosphere approximately five miles high and five miles deep) there is a finite amount of natural resources. People have become dependent on elements such as fresh water, timber, plants, soil, and ore, which are processed into necessary pieces of the human environment.

2. Given the laws of thermodynamics, energy cannot be created or destroyed. The resources that have been allotted to manage existence are contained in the ecosystem.

3. All forms of energy tends to seek equilibrium and therefore disperse. For example, water falls from the sky, settles on plants, and then percolates into the soil to reach the subterranean aquifer. Toxic liquids, released by humans and exposed to the soil, will equally disperse and eventually reach the same underground reservoir. The fresh water aquifer, now contaminated, is no longer a useful natural resource.

There is a need to focus on the preservation of beneficial natural elements and diminish or extinguish natural resources contaminated with toxins and destructive human practices.

There are many credos for environmental responsibility. One, *The Natural Step,* was organized by scientists, designers, and environmentalists in 1996. They were concerned with the preservation of the thin layer that supports human life in a small zone on the earth's surface: the ecosphere (five miles of the earth's crust) and the biosphere (five miles into the troposphere of the atmosphere).

Their principles are as summarized as follows:

1. Substance from the earth's crust must not systemically increase in the ecosphere.

   Elements from the earth such as fossil fuel, ores, timber, etc., must not be extracted from the earth at a greater rate than they can be replenished.

2. Substances that are manufactured must not systemically increase in the ecosphere.

    Manufactured materials cannot be produced at a faster rate than they can be integrated back into nature.

3. The productivity and diversity of nature must not be systemically diminished.

    This means that people must protect and preserve the variety of living organisms that now exist.

4. In recognition of the first three conditions, there must be a fair and efficient use of resources to meet human needs.

    This means that human needs must be met in the most environmentally sensitive way possible.

    Buildings consume at least 40 percent of the world's energy. Thus they account for about a third of the world's emissions of heat-trapping carbon dioxide from fossil fuel burning, and two-fifths of acid rain-causing carbon dioxide and nitrogen oxides.*

The built environment has a monumental impact on the use of materials and fuels to create shelter for human beings. The decisions about the amount and type of materials and systems that are employed in the building process have an enormous impact on the future use of natural resources. Architects can affect and guide those decisions of design to influence the needs of sustainability and environmental sensitivity.

---

* *Sources: David Malin Roodman and Nicholas Lessen, "Building Revolution: How Ecology and Health Concerns are Transforming Construction."* Worldwatch Paper *124 (Washington DC, Worldwatch Institute, 1995);*

Sandra Mendler & William Odell, The HOK Guidebook to Sustainable Design, *(New York: John Wiley & Sons, Inc., 2000).*

SUSTAINABLE SITE PLANNING AND DESIGN

Most architectural projects involve the understanding of the design within the context of the larger scale neighborhood, community, or urban area in which the project is placed.

If the building will be influenced by sustainable design principles, its context and site should be equally sensitive to environmental planning principles.

Sustainable design encourages a re-examination of the principles of planning to include a more environmentally sensitive approach. Whether it is called Smart Grow, sustainable design, or environmentally sensitive development practice, these planning approaches have several principles in common.

Site Selection

The selection of a site is influenced by many factors including cost, adjacency to utilities, transportation, building type, zoning, and neighborhood compatibility. In addition to these factors, there are sustainable design standards that should be added to the matrix of site selection decisions:

■ **Adjacency to public transportation**

    If possible, projects that allow residents or employees access to public transportation are preferred. Allowing the building occupants the option of traveling by public transit may decrease the parking requirements, increase the pool of potential employees and remove the stress and expense of commuting by car.

■ **Flood plains**

    In general, local and national governments hope to remove buildings from the level of the 100-year floodplain. This can be accomplished by either raising the building at least one foot above the 100-year elevation or locating the project entirely out of the 100-year floodplain.

This approach reduces the possibility of damage from flood waters, and possible damage to downstream structures hit by the overfilled capacity of the floodplain.

■ **Erosion, fire, and landslides**

Some ecosystems are naturally prone to fire and erosion cycles. Areas such as high slope, chaparral ecologies are prone to fires and mud slides. Building in such zones is hazardous and damaging to the ecosystem and should be avoided.

■ **Sites with high slope or agricultural use**

Sites with high slopes are difficult building sites and may disturb ecosystems, which may lead to erosion and topsoil loss. Similarly, sites with fertile topsoil conditions—prime agricultural sites—should be preserved for crops, wildlife, and plant material, not building development.

■ **Solar orientation, wind patterns**

Orienting the building with the long axis generally east west and fenestration primarily facing south may have a strong impact on solar harvesting potential. In addition, protecting the building with earth forms and tree lines may reduce the heat loss in the winter and diminish summer heat gain.

■ **Landscape site conditions**

The location of dense, coniferous trees on the elevation against the prevailing wind (usually west or northwest) may decrease heat loss due to infiltration and wind chill factor. Sites with deciduous shade trees can reduce summer solar gain if positioned properly on the south and west elevations of the buildings.

Alternative Transportation

Sites that are near facilities that allow several transportation options should be encouraged. Alternate transportation includes public transportation (trains, buses, and vans); bicycling amenities (bike paths, shelters, ramps, and overpasses); carpool opportunities that may also connect with mass transit; and provisions for alternate, more environmentally sensitive fuel options such as electricity or hydrogen.

Reduction of Site Disturbance

Site selection should conserve natural areas, and restore wildlife habitat and ecologically damaged areas. In some areas of the United States, less than 2 percent of the original vegetation remains. Natural areas provide a visual and physical barrier between high activity zones. Additionally, these natural areas are aesthetic and psychological refuges for humans and wildlife.

Storm Water Management

There are several ways by which reduced disruption of natural water courses (rivers, streams and natural drainage swales) may be achieved:

■ Provide on-site infiltration of contaminants (especially petrochemicals) from entering the main waterways. Drainage designs that use swales filled with wetland vegetation is a natural filtration technique especially useful in parking and large grass areas.

■ Reduce impermeable surface and allow local aquifer recharge instead of runoff to waterways.

■ Encourage groundwater recharge.

Ecologically Sensitive Landscaping

The selection of indigenous plant material, contouring the land, and proper positioning of shade trees may have a positive effect on the landscape appearance, maintenance cost, and ecological balance. The following are some basic sustainable landscape techniques:

- Install indigenous plant material, which is usually less expensive, to ensure durability (being originally intended for that climate) and lower maintenance (usually less watering and fertilizer).

- Locate shade trees and plants over dark surfaces to reduce the "heat island effect" of surfaces (such as parking lots, cars, walkways) that will otherwise absorb direct solar radiation and retransmit it to the atmosphere.

- Replace lawns with natural grasses. Lawns require heavy maintenance including watering, fertilizer, and mowing. Sustainable design encourages indigenous plant material that is aesthetically compelling but far less ecologically disruptive.

- In dry climates, encourage xeriscaping (plant materials adapted to dry and desert climates); encourage higher efficiency irrigation technologies including drip irrigation, rainwater recapture, and gray water reuse. High efficiency irrigation uses less water because it supplies water directly to the plant's root areas.

Reduction of Light Pollution

Lighting of site conditions, either the buildings or landscaping, should not transgress the property and not shine into the atmosphere. Such practice is wasteful and irritating to the inhabitants of surrounding properties. All site lighting should be directed downward to avoid "light pollution."

Open Space Preservation

The quality of residential and commercial life benefits from opportunities to recreate and experience open-space areas. These parks, wildlife refuges, easements, bike paths, wetlands, or play lots are amenities that are necessary for any development.

In addition to the aforementioned water management principles, the following are principles of design and planning that will help increase open-space preservation:

- **Promote in-fill development** that is compact and contiguous to existing infrastructure and public transportation opportunities.

  In-fill development may take advantage of already disturbed land without impinging on existing natural and agricultural land.

  In certain cases, in-fill or redevelopment may take advantage of existing rather than new infrastructure.

- **Promote development that protects natural resources** and provides buffers between natural and intensive use areas.

  First, the natural areas (wetlands, wildlife habitats, water bodies, or flood plains) in the community in which the design is planned should be identified.

  Second, the architect and planners should provide a design that protects and enhances the natural areas. The areas may be used partly for recreation, parks, natural habitats, and environmental education.

  Third, the design should provide natural buffers (such as woodlands and grasslands) between sensitive natural areas and areas of intense use (factories, commercial districts, housing). These buffers may offer visual, olfactory, and auditory protection between areas of differing intensity.

Fourth, linkages should be provided between natural areas. Isolated islands of natural open space violate habitat boundaries and make the natural zones seem like captive preserves rather than a restoration or preservation of natural conditions.

Fifth, the links between natural areas may be used for walking, hiking, or biking, but should be constructed of permeable and biodegradable material. In addition, the links may augment natural systems such as water flow and drainage, habitat migration patterns, or flood plain conditions.

■ **Establish procedures that ensure the ongoing management of the natural areas** as part of a strategy of sustainable development.

Without human intervention, natural lands are completely sustainable. Cycles of growth and change including destruction by fire, wind, or flood have been occurring for millions of years. The plants and wildlife have adapted to these cycles to create a balanced ecosystem.

Human intervention has changed the balance of the ecosystem. With the relatively recent introduction of nearby human activities, the natural cycle of an ecosystem's growth, destruction, and rebirth is not possible.

Human settlement will not tolerate a fire that destroys thousands of acres only to liberate plant material that reblooms into another natural cycle.

The coexistence of human and natural ecosystems demands a different approach to design. This is the essence of sustainable design practices, a new approach that understands and reflects the needs of both natural and human communities.

AHWAHNEE PRINCIPLES

In 1991, in the Ahwahnee Hotel in Yosemite National Park, a group of architects, planners, and community leaders met to present community principles that express new, sustainable planning ideas. These principles are summarized below.

Preamble

Existing patterns of urban and suburban development seriously impair our quality of life. The symptoms are: more congestion and air pollution resulting from our increased dependence on automobiles, the loss of precious open space, the need for costly improvements to roads and public services, the inequitable distribution of economic resources, and the loss of a sense of community. By drawing upon the best from the past and the present, we can plan communities that will more successfully serve the needs of those who live and work within them. Such planning should adhere to certain fundamental principles.

Community Principles

1. All planning should be in the form of complete and integrated communities containing housing, shops, workplaces, schools, parks, and civic facilities, essential to the daily life of the residents.

2. Community size should be designed so that housing, jobs, daily needs, and other activities are within easy walking distance of each other.

3. As many activities as possible should be located within easy walking distance of transit stops.

4. A community should contain a diversity of housing types to enable citizens from a wide

range of economic levels and age groups to live within its boundaries.

5. Businesses within the community should provide a range of job types for the community's residents.

6. The location and character of the community should be consistent with a larger transit network.

7. The community should have a center focus that combines commercial, civic, cultural, and recreational uses.

8. The community should contain an ample supply of specialized open space in the form of squares, greens, and parks, whose frequent use is encouraged through placement and design.

9. Public spaces should be designed to encourage the attention and presence of people at all hours of the day and night.

10. Each community or cluster of communities should have a well-defined edge, such as agricultural greenbelts or wildlife corridors, permanently protected from development.

11. Streets, pedestrian paths, and bike paths should contribute to a system of fully connected and interesting routes to all destinations. Their design should encourage pedestrian and bicycle use by being small and spatially defined by buildings, trees, and lighting, and by discouraging high speed traffic.

12. Wherever possible, the natural terrain, drainage, and vegetation of the community should be preserved with superior examples contained within parks or greenbelts.

13. The community design should help conserve resources and minimize waste.

14. Communities should provide for the efficient use of water through the use of natural drainage, drought tolerant landscaping, and recycling.

15. The street orientation, the placement of buildings, and the use of shading should contribute to the energy efficiency of the community.

Regional Principles

1. The regional land-use planning structure should be integrated within a larger transportation network built around transit rather than freeways.

2. Regions should be bounded by and provide a continuous system of greenbelt/wildlife corridors to be determined by natural conditions.

3. Regional institutions and services (government, stadiums, museums, and so forth) should be located in the urban core.

4. Materials and methods of construction should be specific to the region, exhibiting a continuity of history and culture and compatibility with the climate to encourage the development of local character and community identity.

Implementation Principles

1. The general plan should be updated to incorporate the above principles.

2. Rather than allowing developer-initiated, piecemeal development, local governments should take charge of the planning process. General plans should designate where new growth, in-fill, or redevelopment will be allowed to occur.

3. Prior to any development, a specific plan should be prepared based on these planning principles.

4. Plans should be developed through an open process and participants in the process should be provided visual models of all planning principles.

Source: Local Government Commission's Center for Livable Communities, http://lgc.org/clc/.

USGBC—U.S. GREEN BUILDING COUNCIL

Incorporated as a nonprofit trade association in 1993, the U.S. Green Building Council (USGBC) was founded with a mission "to promote buildings that are environmentally responsible, profitable and healthy places to live and work." It is formed of leaders from across the building industry who head a national consensus for producing a new generation of buildings that deliver high performance inside and out.

The core of the USGBC's work is the creation of the Leadership in Energy and Environmental Design (LEED) green building rating system. LEED provides a complete framework for assessing building performance and meeting sustainability goals. Based on well-founded scientific standards, LEED emphasizes state of the art strategies for sustainable site development, water savings, energy efficiency, materials selection, and indoor environmental quality. LEED recognizes achievements and promotes expertise in green building through a comprehensive system offering project certification, professional accreditation, training, and practical resources.

USGBC committees are actively collaborating on new and existing LEED standards, including a standard for homes, neighborhood development, and commercial interiors.

Their Web site is: *www.usgbc.org.*

ARCHITECTURAL PROCESS

After the planning process has been concluded, and the site has been selected, the architectural team will begin to focus on the project, including the project's buildings and related infrastructure.

Traditionally, the architect is faced with four components to every design decision: cost, function, aesthetics, and time. The new paradigm adds *sustainability* to this list.

The ingredients of the normal process have been discussed previously, but the new ingredient, sustainability, changes the meaning of all these pieces of the architectural process.

Cost

As architects put together budgets for their clients, they are always concerned with the first costs of the design components—the initial cost to purchase and install the design element.

Sustainable design has made the economic decision process more holistic. The decision to select a design element (such as a window, door, flooring, exterior cladding, or mechanical system) is now concerned with the "life cycle" cost of the design.

Life-Cycle Costing

Life-cycle costing is concerned not only with the first cost, but the operating, maintenance, periodic replacement, and residual value of the design element.

For example, two light fixtures (A and B) might have different first cost: Fixture A has a 10 percent more expensive first cost than B. But when the cost of operation (the lamps use far less energy per lumen output) and the cost of replacement (the bulbs of A last 50 percent longer than the bulbs of Fixture B) is evaluated, Fixture A has a far better life-cycle cost and should be selected.

In this kind of comparison, the life-cycle cost may be persuasive; the extra cost of Fixture A may be recovered in less than two years due to more efficient operation and replacement savings. In this situation the architect justified Fixture A to the owner, who benefits from a more energy efficient lighting that continues to save the owner operating costs for the life of the building.

Matrix Costing

While designing a typical project, the architect faces numerous alternate decisions, a process that may be both intriguing and complex.

Sustainable design adds an ingredient to the matrix of decisions that may actually help the composition.

For example, decisions that allow the improved efficiency of the building envelope, light fixtures, and equipment may permit the architect to allow the engineer to reduce the size of the HVAC system, resulting in a budgetary trade-off. The extra cost of the improved envelope may be economically balanced by the diminished cost of the mechanical system.

This type of economic analysis, which evaluates cost elements in a broad matrix of interaction, is a very valuable architectural skill. The ability to understand the interaction between different building systems in a creative and organized fashion can differentiate an excellent from a simply adequate architectural design.

Function

Functionality is one of the primary standards of architectural design. If the building does not perform according to the client's needs, then the building design has failed.

Years ago, the design element could perform at the highest level regardless of its impact on the environment or energy use. The fact that many industrial and residential buildings are operating in 2003 much more efficiently than 1960 is evidence that the building design and construction profession is learning how to tune buildings to

a higher degree of energy operation. But, with diminishing natural resources and increasing pollution of the environment, even more efficient design is necessary.

Today, architects will include sustainability in the selection of optimal functional design components.

For example, a roof system must be able to withstand a variety of weather conditions, be warranted to be durable a minimum of years, be able to be applied in a range of weather conditions, and have a surface with reflectivity that does not add to the urban heat effect.

Time

The schedule of a project is always a difficult part of the reality of the design process. Time is a constraint that forces a systematic and progressive evaluation of the design components.

The sustainable component of the architectural process may add to the amount of time the architect will spend on the research for the project.

The architect may spend more time on a sustainable design with the result being a more integrated, sustainable project.

Aesthetics

The aesthetic of a project is the combination of the artistry of the architect and the requirements of the project.

Sustainable design has the reputation of emphasizing function and cost over beauty and appeal.

It is the architect's responsibility to keep all the design tools in balance. A project without aesthetic consideration will fail the client, its user, and the potential client who may be deciding between the normal design and one that considers a broader, integrated, sustainable approach.

Sustainability

The fifth point is a new component that leads to a new approach to the design process.

Sustainable designs should have five goals:
1. Use less
2. Recycle components
3. Use easily recycled components
4. Use fully biodegradable components
5. Do not deplete natural resources necessary for the health of future generations

STANDARDS FOR EVALUATION

How can we objectively evaluate the quality of a sustainable project?

The architect is faced with responding to many standards and regulations in the course of assembling a design. Building codes, life safety standards, fire code, zoning regulations, and health and sanitary regulations are some of the many municipal, state, and federal standards that an architect must evaluate in the course of any project.

Sustainability is a new filter for the design process and there are several organizations that have offered checklists for evaluating the inclusion of environmentally sensitive elements into the project.

One of the measures of performance is LEED (Leadership in Energy & Environmental Design), which is sponsored by the USGBC (U.S. Green Building Council). This standard was developed in the 1990s by a consortium of

building owners, architects, suppliers, engineers, contractors, and governmental agencies.

The goal of LEED and similar environmental design standards is to introduce new sustainable approaches and technologies to the construction industry. LEED is a voluntary environmental rating system that is organized into six categories:
1. Sustainable sites
2. Water efficiency
3. Energy and atmosphere
4. Materials and resources
5. Indoor air quality
6. Innovation and design practice

LEED covers the range of architectural decisions, including site design, water usage, energy conservation and production, indoor air quality, building materials, natural lighting, views of the outdoors, and innovative design components.

The LEED point award matrix is a mixture of teaching, persuasion, example, and incentive. It is good checklist for the entire project team to evaluate the quality of sustainable design decisions for the complete project—from initial planning through final construction, maintenance, and training procedures.

These categories combine *prerequisites* (basic sustainable practices such as building commissioning, plans for erosion control, or meeting minimum indoor air quality standards) with optional *credits* (water use reduction, heat island reduction, or measures of material recycled content).

Most of the credits are performance based—solutions based on system performances against an established standard such as American Society of Heating, Refrigeration and Air Conditioning

Engineers (ASHRAE). ASHRAE has created one of most widely recognized standards of energy design that is used by mechanical engineers and architects.

For example, one credit (under the Energy and Resources category) is "Optimize Energy Performance."

The number of points for this credit depend on how the architectural and engineering team can optimize the design of the building's energy systems against the ASHRAE 90.1 standards.

The possible design solutions include optimizing the heating, cooling, fans, pumps, water, and interior lighting systems.

In the graduated point matrix for a new building, if the team improves the performance (against ASHRAE standards) by 15 percent they receive one point and if they manage to improve by 60 percent they receive ten points.

LEED describes suggested results but allows the architectural team to find a variety of solutions. The LEED certification awards range from Bronze at 40 percent compliance to Platinum at 81 percent compliance. The LEED certification is innovative and rigorous, and currently there are fewer than a half dozen platinum buildings in the United States.

THE SUSTAINABLE DESIGN PROCESS

Is a sustainable design organized and implemented differently from a conventional design?

The Design Team

What kind of design team is necessary for a sustainable project?

The scope of a sustainable design invites an expanded team approach, which may include the following:

■ Architects or engineers (structural, MEP) with energy modeling experience

■ A landscape architect with a specialty in native plant material

■ A commissioning expert (if LEED employed)

■ An engineer/architect with building modeling experience

The design team for a sustainably designed project tends to have a larger pool of talent than a typical architectural project. Wetlands scientists, energy efficient lighting consultants, native plant experts, or commissioning engineers are examples of the additional talent that may be added to sustainable design projects.

As with any architectural design, there is a hierarchy of design goals:

■ *Initial imperatives* such as budget, timing, image, and program necessities

■ *Subjective goals* such as a functionally improved and more pleasing work environment, pleasing color schemes, and landscaping that complements the architecture

■ *Specific goals* such as more open space, more natural light, less water usage, and adjacency to public transportation

And with the inclusion of sustainability there may be additional goals:

■ *Initiatives that are specific to sustainability* such as fewer toxins brought into the space, daylighting in all spaces with people occupancies, less overall energy consumed, less water usage, adjacency to public transportation, and improved indoor air quality

■ Desire to exceed existing standards such as ASHRAE, USGBC, or American Planning Association (APA)

RESEARCH AND EDUCATION

Is additional education and research necessary for a sustainable project?

Yes. Innovative HVAC systems, durable yet non-toxic materials, recycled materials, recyclable materials, native plant material, energy efficient lighting, and controls are examples of design components that are not normally designed and installed by general contractors and architectural consultants on typical projects.

Education of the Client

Sustainable design requires a new way of examining the architectural design process. Concepts such as life-cycle costing, recycled versus recyclable materials; non-VOC (volatile organic compounds) substances; daylighting; and alternate energy sources are among the several new concepts that the architect should discuss with the client before the design process commences.

It is critical that the client understands the sustainable process and is sympathetic to its potential economic and environmental benefits.

Education of the Project Team

Once the project has been assigned to an architect, but before the design process begins, the project team (architect, engineer, contractor, consultants, and owner) should assemble and discuss the project scope and objectives with all the project team members.

Establishing Project Goals

Among the many items included in the scope of work (including the extent of work, program elements, budget, and schedule) are the objectives for sustainable design.

For example, the architect and owner might establish goals for several environmental areas such as:

■ X percent reduction of energy usage from the established norm (see "Benchmarking" later in this section)

■ Improved lighting (less energy used and more efficient dispersal of indirect light with less glare)

■ Nontoxic and low VOC paint and finishes

■ Increased recycled content in materials such as carpeting, gypsum wallboard, ceiling tiles, metal studs, and millwork

■ High-efficiency (energy star) appliances

■ Wood elements are all certified wood products

■ Daylighting in all work/occupied spaces

As the leader of the project team, it is the architect's responsibility to include sustainable goals with the rest of the project scope of work.

A detailed explanation of the benefits of these sustainable design elements to all of the project team will ensure that they fully understand the design potential and economic implications of these concepts.

Verify Extent of Work

Sustainable design involves a more comprehensive approach to pre-project planning.

The LEED certification process will require record keeping and verification of the source of materials—a process that is beyond the normal design and construction work. For purposes of selecting a contractor and consultants, the team should be briefed on these additional obligations.

For example, the demolition process (if LEED certified) will require verification that materials have been sorted and delivered to an approved recycling organization. By contrast, the normal demolition process does not require recycling or verification that each material is sorted by type.

Clearly establishing the extent and type of effort required for each member of the sustainable design team is critical. The extent and type of effort will affect each member's ability to participate and their fees for services and construction work.

Energy and Optimization Modeling

Building shape, orientation, fenestration location, roof color, envelope configuration, and HVAC system efficiency are some of the variables in sustainable design projects that can be fine tuned with DOE-2 (U.S. Department of Energy's building analysis program) and other computer energy modeling programs.

The "fine tuning" of a project's energy components is one of the elements in the architect's design matrix that affects the final appearance, cost, and performance of the final design.

Energy modeling will not govern the final design. Issues such as compatible scale, color, texture, and functionality are still part of the architect's palette. But energy modeling is one additional factor that the architect will employ as part of the "best practices" approach to architecture.

In addition, modeling can assist in the cost analysis of a project. The fact that the modeling program is interactive helps the architect simultaneously adjust design elements to demonstrate alternate energy efficient solutions.

For example, energy modeling might allow the architect to demonstrate to the team that a

more durable, aesthetically pleasing, and energy efficient building skin could be economically justified by reducing the size and cost of the mechanical system.

The ability to visually and numerically quantify the efficacy of trading certain design elements may be an effective tool for the architect when discussing the building design with the consultants and owner.

The Bid and Specification Process

The requirements of a sustainable design will often vary from a normal project.

For example, the millwork section of bid documents will normally specify the finish material, configuration of the design and methods of attachment, delivery, and installation. But the requirement of non-VOC glues and non-VOC substrate may confuse a potential bidder and cause that bidder to increase the bid price unnecessarily.

To facilitate the bidding and construction process, the architect should include the following:

- Simple definitions of sustainable elements—for example, what "VOC," "certified" wood product, or "daylighting" mean

- Explanations of specific characteristics of sustainable elements—for example, specifically state the standard that must be met (for example, Green Label Testing Program Limits, carpet's total VOC limit, that is, formaldehyde 0.05 (mg/m2)

- References of specific regulatory agency's information (name, address, e-mail, phone, and so on)—for example, the Carpet and Rug Institute, *www.carpet-rup.com*, (800) 555-8846

- Examples of suppliers that could meet the sustainable standards indicated—in the case of sustainable products, there are at least two approaches to a list of suppliers for products:

1. Limit the installer to three to five suppliers of a product that is known to satisfy the sustainable design specifications.

   This approach assures the architect that the product will meet specified standards.

   (Note, however, that with the constantly changing nature of the emerging sustainable design market, a limited list could limit competition and the diversity of creative alternatives.)

2. Identify a list of qualified suppliers, but permit the bidder/contractor to submit alternative suppliers who satisfied the sustainable design criteria. This approach creates a more competitive environment, but it will require more effort of the architect to properly review and qualify the bids.

Changes and Substitutions

Every project is faced with the reality of time and budgetary pressures. And, in those instances, there may be situations when one product or design element may not be available in the form originally specified.

Sustainable designed projects require more stringent architectural supervision to ensure that original design standards are met. For example, in the rush to project completion the installer may claim that paints used for "touch up" of damaged areas are so small that they may be installed with normal, higher-VOC paints. This minor transgression might jeopardize the integrity of the project and the ability to receive certification for LEED credits in certain areas.

ENERGY EVALUATION

In the climates of North America, buildings need some form of purchased energy (electricity, natural gas, oil) in order to operate. The architect works with his or her team to design strategies that may reduce the amount of purchased energy, reduce operating costs, and reduce the nation's dependence on imported fossil fuels.

The following are some design strategies that the sustainable design approach might employ to improve a building's energy performance. These elements are listed and briefly described.

Solar Design

Solar design is the age-old system of using sunlight or solar radiation to supply a portion of the building's heat energy. By a combination of techniques such as window and skylight design, location of internal thermal mass, and internal organization of the building's functions, solar design may replace some of the fossil fuel needed for heating and cooling buildings.

Passive solar systems is a category of solar design. Passive solar systems are those systems that permit solar radiation to fall on areas of the building that benefit from the seasonal energy conditions of the structure.

For example, some North American buildings are designed to reduce solar radiation gains from sunlight in the summer. Passive solar design relies on inherent qualities of the building's fenestration, massing, and orientation to capture sunlight.

Passive solar systems are usually categorized into direct or indirect gain systems.

Direct gain systems, as the title implies, are those systems that allow solar radiation to flow directly into the space needing heat. A process commonly known as the "greenhouse effect" allows much of the sunlight that passes through the glass of the fenestration to be retained in the material it strikes (stone, concrete, wood, etc.) inside the building. Thus, south facing windows allow solar radiation to be directly gained and used inside the building.

Indirect gain systems operate when the sunlight first strikes a thermal mass that is located between the sun and the space. The sunlight absorbed by the mass is converted to thermal energy (heat) and then transferred into the living space.

There are basically two types of indirect gain systems: thermal storage walls and roof ponds. The difference is essentially the location—roof verses wall materials.

Passive solar design might employ several architectural strategies to facilitate the design:

1. *Architectural sun control devices.* Overhangs or shading devices that have been designed to permit winter solar radiation from entering the building interior while blocking the higher angled, summer solar radiation from entering the building. Deciduous trees often perform the same function of permitting winter sunlight to enter and blocking much of the summer solar radiation with branches and leaves. Other examples include shutters; vertical projections or fins; awnings; trellises (especially with shading vegetation); and sunscreens (some with PV panels that both gather sunlight to convert into electricity and shade unwanted radiation from interior space in the warm months).

2. *Light-colored roof systems.* Light-colored roofing materials reflect sunlight and reduce the amount of radiation that is absorbed through the roof into the interior space. Colors with higher reflectance (albedo) factors are preferred. For example, some cities in the United States require roof materials to have a minimum albedo rating of .65 (65 percent of the solar radiation is reflected back into the atmosphere). The urban heat island effect, caused by roofs, roads, and parking areas that absorb solar radiation during the day and retransmit the stored heat during the afternoon and evening, can be modified with light-colored roof systems.

   By designing these surfaces with light-colored and reflective material, the amount of heat energy stored in these materials is diminished and the urban heat island effect is reduced. Grass or vegetated roof areas have good insulating value and may also reduce the urban heat island effect and provide cooling through evapotranspiration.

3. *Optimized building glazing systems.* Orientation, light transmittance factors, and U-value are all factors architects consider in selecting glazing. Glass that is low-E (emissivity) is desirable because it is coated with a material that allows a maximum amount of sunlight to be transferred through the glass and not reflected back into the atmosphere.

Lighting

The illumination of the interior of a sustainably designed building requires a holistic approach that balances the use of artificial and natural lighting sources.

Daylighting

Properly filtered and controlled solar radiation may provide a valuable source of illumination to a building interior. This process is called "daylighting" (simply having properly designed fenestration that allows natural sunlight to replace or dramatically reduce the need for artificial lighting).

Because unwanted sunlight (particularly in summer months) can also add to the internal heat load of a building, the architect must be careful to allow only beneficial sunlight and reduce unwanted solar heat gain. There are several techniques for controlling daylighting:

1. Overhangs, fins, and other architectural shading devices

2. Sawtooth (not bubble) skylight design, which allows the glass to face north for illumination, not south for solar heat gain

3. Interior window shading devices, which allow solar gain during cool months, and the blocking of solar radiation during the warmer seasons

4. Light shelves, which permit the daylight to reflect off the ceiling and penetrate farther into the interior without affecting views outside

Higher Efficiency Light Fixtures

In addition to a daylighting strategy, light fixtures that are more efficiently designed reduce energy cost and increase comfort, such as the following:

- Fixtures that use fluorescent or HID lamps, which provide more illumination per watt than incandescent lighting.

- Fixtures that are designed to diffuse or bounce the illumination off the ceilings or internal reflectors, which are more efficient; cause less glare (especially in an environment with computer monitors); and save operating costs.

- Fixtures that have higher efficiency (T-8) fluorescent bulbs, which produce more

lumens per watt and thereby diminish the heat generated by lighting.

- Fixtures that offer dimming or multiple switching capability, which permit the architect a more energy efficient lighting design. Dimming or multiple switching fixtures allow the architect to design lighting patterns that blend nicely with daylighting opportunities. For example, an office with perimeter fenestration allows daylighting supplemented with overhead lighting that can be dimmed or reduced. The interior spaces, which are too far from the perimeter for daylighting, may be controlled with switches or dimmers that allow relatively higher levels of illumination. The result is an even illumination pattern, which saves on artificial lighting costs, by relying on daylighting at the perimeter.

- Fixtures that use higher efficiency lamps such as fluorescent, high intensity discharge (HID) sulfur lighting (exterior only).

- Fluorescent fixtures that use high efficiency electronic ballasts.

Additionally, the architect may avoid less efficient incandescent lighting where possible; install task lighting to supplement diffused ambient lighting; and install LED (light emitting diode) lighting for exit signs. LED lighting lasts longer than incandescent and is far less expensive to operate.

Lighting Sensors and Monitors

Where possible, lighting costs can be diminished by installing light monitors that sense occupancy conditions. As long as the room contains people, the lights will remain on. If people leave, the sensor will wait for a few minutes, then shut off all the lighting in the room.

Lighting sensors can be designed to operate with a preference for motion, heat (from people), or desired time of occupancy.

Lighting Models

Computer lighting models are one option that allows the architect to simulate the levels of sunlight that penetrate into a building design, depending on the building location, varying times of year, fenestration orientation, and design.

By incrementally altering fenestration (skylights, windows, or light transport systems) and the artificial lighting system, the architect may optimize the daylighting and artificial lighting systems for the building.

Benchmarking

The U.S. Department of Energy provides "benchmark" information of total energy consumption in BTUs/SF for various kinds of buildings in the United States. These standards, or benchmarks, can be useful in the measuring of energy efficiency standards for various types of buildings:

For example:

- Average for all office buildings (pre-1990) 104.2
- Average for all office buildings (1990–1992) 87.4
- Average for all educational buildings (pre-1900) 87.2
- Average for all educational buildings (1990–1992) 57.1
- Average for all laboratory buildings (pre-1990) 319.2
- Average for all health care buildings (pre-1990) 218.5

Source: U.S. Department of Energy, Commercial Building Energy Consumption and Expenditures.

Benchmarking is a good way to alert the design team to the base energy standards for their design. It's a good place to start and ultimately a standard to beat. And, one can see from some of the comparisons (office and educational buildings), that some energy efficiency is occurring.

COMMISSIONING

Commissioning is an organized process to ensure that all building systems perform interactively according to the intent of the architectural and engineering design, and the owner's operating needs.

Commissioning usually includes all HVAC and MEP systems, controls, ductworks and pipe insulation, renewable and alternate technologies, life safety systems, lighting controls and daylighting systems, and any thermal storage systems. Commissioning also verifies the proper operation of architectural elements such as the building envelope, vapor and infiltration control, and gaskets and sealant used to control water infiltration.

Commissioning is a process required for LEED certification, but is a recommended procedure for any building involved with sustainable design procedures.*

* Source: Commissioning Requirements for LEED Green Building Rating, Version 8. February 5. 1999; Sandra Mendler and William Odell, The HOK Guidebook to Sustainable Design, New York, John Wiley & Sons, Inc.: 2000, p. 71.

INNOVATIVE TECHNOLOGIES

Besides the aforementioned issues of solar design, improved lighting systems, improved HVAC systems, and improved building massing and envelope design, there are several "innovative technologies" that the architect can offer to the project team for consideration.

Ground Water Aquifer Cooling and Heating (AETS)

One alternative to full air-conditioning with chillers, which make heavy demands on electricity, is the aquifer thermal energy storage, which uses the differential thermal energy in water from an underground well to cool a building during summer and heat a building in the winter.

This is an efficient, relatively low-cost system, but it may require approval from the local environmental authority before installation.

Geothermal Energy

Where appropriate, heat contained within the earth's surface causes macro-geological events (such as underground geothermal springs or lava formations) that may be tapped to produce heat for adjacent structures.

In select locations this heat energy can be transferred and conveyed to supplement a building's heating demand.

Wind Turbines

Small-scale wind machines used to generate electricity can be mounted on buildings or in open space nearby. These systems share several advantages:
- Relatively cost-effective
- Tested and established technology
- Systematic started-up
- Relatively high output

These systems share several disadvantages:
- Need a relatively high mast
- Require substantial structural support
- Present potential for noise pollution
- Visually intrusive

Photovoltaic (PV) Systems

The basis of the PV systems is the concept that electricity is produced from solar energy when photons or particles of light are absorbed by semiconductors.

Most PV systems are mounted to the building (either on the roof or as shading devices above fenestration). Currently, PV systems are not cost effective. But with promised government subsidy necessary to achieve an economy of scale, PVs may be a viable method of electrical production in the United States, Japan, and Germany in the near future.

Fuel Cells

Even though Sir William Grove invented the technology for the fuel cell in 1839, it has only recently been recognized as a potential power source for the future. The fuel cell claims to be the bridge between the hydrocarbon economy and the hydrogen-based society.

Fuel cells are electrochemical devices that generate direct current (DC) electricity similar to batteries. But, unlike batteries, they require a continual input of hydrogen-rich fuel. In essence, fuel cells are reactors that combine hydrogen and oxygen to produce electricity, heat, and water. They are clean, quiet, and emit no pollution when fed directly with hydrogen.

At the moment, fuel cell technology is still not cost effective for the commercial building market. Still, there seems to be a general feeling that hydrogen-based energy reactors will soon be an optional energy source.

Biogas

Biogas is produced through a process that converts biomass, such as rapid-rotation crops and selected farm and animal waste, to a gas that can fuel a gas turbine. This conversion process occurs through anaerobic digestion—the conversion of biomass to gas by organisms (like bacteria) in an oxygen-free environment.

Biogas has several advantages: it has relatively high energy production; it lends itself to both heat and power production; it creates almost zero carbon dioxide emissions; it virtually eliminates noxious odors and methane emissions; and it protects ground water and reduces the landfill burden.

Small-Scale Hydro

Harnessing the energy from moving water is one of the oldest energy production systems in the world. In some locations, small-scale hydro power is a efficient and clean source of energy and is devoid of environmental penalties associated with large scale hydro projects. It allows small scale, local energy production, with relatively low cost.

Ice Storage Cooling Systems

One of the problems for energy supply companies is that the highest demand for electricity often coincides with the highest cooling demand.

The utilities would prefer to "flatten the curve" (to even out or flatten the measure of average daily energy demand). The fewer the number of peaks (high points of energy demand), the less the utilities have to bolster their power supply with expensive, supplemental fuels.

One way to reduce this peaking problem is to supplement a building's cooling capacity with an ice storage system.

An ice storage system has three components: a tank with liquid storage balls, a heat exchanger, and a compressor for cooling. The essence of the ice storage system is that the chilling and freezing of the ice balls occurs at night (when the cost of energy is lower due to lower demand). During the day, the cool temperatures, stored in the ice, are transmitted into the building's cooling system.*

CONCLUSION

The knowledge of environmental systems has become essential to the architect's design palette. Buildings that take advantage of natural systems such as sun, wind, rain, ground water, topography, and climate are more elegant solutions. Architectural designs that incorporate natural systems, in conjunction with contemporary technologies, are in the tradition of architects providing spatial solutions with the most innovative contemporary thinking available.

Buildings with this approach operate more efficiently, integrate effectively into their local environment, and tend to produce spaces that are more pleasing to work and live. Knowledge of integrated or holistic design principals is not a limitation but another set of tools to produce humane, efficient, healthy, and aesthetically compelling architecture.

Source: Peter F. Smith, Sustainability at the Cutting Edge, Jordan Hill, Oxford: Architectural Press, an Imprint of Elsevier Science, 2003.

LESSON 1 QUIZ

1. Sustainable design is primarily concerned with which of the following issues?
   - I. Economics
   - II. Aesthetics
   - III. Environment
   - IV. Mechanical systems
   - **A.** III
   - **B.** I, II, and III
   - **C.** I and III
   - **D.** All of the above

2. The *Natural Step* is an approach to the environment that follows which of the following principles?
   - I. The biosphere affecting humans is a relatively stable and resilient zone that includes five miles into the earth's crust and five miles into the troposphere.
   - II. Improved technologies have dramatically increased the number and quantity of available natural resources.
   - III. Toxic substances released into either the sea or atmosphere will only influence areas adjacent to the toxic source.
   - IV. Using building materials that are recycled is an adequate sustainable design approach.
   - **A.** I
   - **B.** II
   - **C.** II and IV
   - **D.** None of the above

3. The planning phase of a sustainably designed architectural project should include which of the following elements?
   - I. Native landscaping that is aesthetically pleasing and functional
   - II. Designing structures in the floodplain that can resist the forces of flood waters
   - III. Consideration of sun orientation, topographic relief, and the scale of adjacent buildings
   - IV. Locating projects within existing neighborhoods that are adjacent to public transportation
   - **A.** I and II
   - **B.** I and III
   - **C.** I, III, and IV
   - **D.** All of the above

4. The Ahwahnee principles include which of the following ideas?
   - I. Communities with only residential use should be relegated to areas outside the central business district.
   - II. Preserved open spaces should be either wildlife habitats or recreational areas.
   - III. Transportation planning should include roads, pedestrian paths, bike paths, and mass transit systems.
   - IV. Job creation and economic diversity is a desired goal.
   - **A.** I
   - **B.** II, III, and IV
   - **C.** III and IV
   - **D.** None of the above

5. Life cycle costing is an economic evaluation of architectural elements that includes which of the following factors?

I. First cost

II. Maintenance and operational costs

III. Repair costs

IV. Replacement cost

A. I

B. II, III, and IV

C. II and IV

D. All of the above

6. LEED, the name of a program that environmentally evaluates sustainable projects, is a checklist that is concerned with which of the following?

I. Indoor air quality

II. Storm water

III. Innovative energy systems

IV. Aesthetic design

A. I

B. I, II, and III

C. II and III

D. All of the above

7. Which of the following is a consultant who might be employed in a sustainable design project?

I. Wetlands engineer

II. Energy commissioner

III. Landscape architect

IV. Energy modeling engineer

A. I

B. I and II

C. II, III, and IV

D. All of the above

8. Sustainable design may require research and education that is beyond a normal architectural project. Which of the following is part of this process?

I. Energy modeling

II. Education of the client

III. Art selection

IV. Selection of energy efficient appliances

A. I and IV

B. I and II

C. I, II, and IV

D. All of the above

9. Sensitivity to the nuances of site conditions is key to sustainable design. Which of the following are site conditions the architect should examine in the design process?

I. Solar orientation

II. Decorative landscaping

III. Scale and style of adjacent structures

IV. Ground water conditions

A. I and II

B. I, III, and IV

C. I and III

D. All of the above

10. Sustainably designed architecture requires attention to which of the following building elements?

I. Solar shading devices

II. Urban heat island effect

III. Increased parking

IV. Fenestration and glazing

A. I, II, and IV

B. I and IV

C. I and II

D. All of the above

WATER SUPPLY AND DRAINAGE SYSTEMS

INTRODUCTION

Among the systems which we take most for granted in our buildings are those that provide clean healthy water, and that remove waste water efficiently, without danger, and without breeding disease. Although we give very little thought to modern plumbing, and accept it as a matter of course, obtaining clean water is a major problem in many parts of the world, and lack of clean water is a significant source of disease. The architect should therefore have an understanding of the basic principles of plumbing systems, and the Architect Registration Examination tests this understanding. Let us begin our discussion by separating supply and waste systems.

Supply systems are those that supply clean, clear and potable water for industrial processes, washing, cooking, or drinking. Supply systems are under pressure, and thus must be sealed. Being under pressure, water supply lines can run vertically in buildings, typically in smaller pipes than those used for waste systems. *Sanitary waste* systems, on the other hand, remove contaminated water, and are generally not under pressure. Therefore great care must be taken to drain them by gravity, and to avoid contaminating other systems. *Storm drains* are somewhat similar in that they must be drained by gravity, and typically require much larger pipes. Each of these will be discussed separately, in detail.

SUPPLY

Water must be clean, clear, and *potable*—that is, suitable for drinking. There are several contaminants that may cause trouble and must therefore be considered.

Acidity

When water falls from the sky as rain, it is usually clear of any mineral content, but slightly *acidic*. Acidity is measured by the pH of the water. Neutral water has a pH of 7. The greater the acidity, the *lower* the number. For example, a pH of 6.9 to 6.0 represents slightly acidic water, and a pH of 5 represents very acidic water. A pH *above* 7 represents a *basic* or *alkaline* solution. The pH scale ranges up to 14, which represents the most alkaline, or "basic" solution. The naturally acidic state of rain water has been exacerbated in some areas by the presence of the by-products of combustion in the air, most notably sulfur and nitrogen compounds, which may combine with the moisture present to form sulfuric (most common) or nitric (less common) acid. This has become a major problem in the northern United States and Canada, where the acidity of many fresh water lakes has risen to the point where few fish, or no fish at all, can survive. This may pose a threat to our water supply at some future date. In any case, the acidity of rain water can cause problems in plumbing systems, since it *corrodes* metal pipes. Furthermore, other contaminants from air pollution may be absorbed into the rain water, which may make it dangerous or unfit for human consumption. For this reason, rain water and *surface runoff* may not be as healthy as water that has been absorbed into the ground and partially filtered of these contaminants.

Hardness

Once water has seeped into the ground, it dissolves minerals present in the ground, most notably *limestone ($CaCO_3$)* and/or *calcium* and/or *magnesium.* This is called *hard* water. This is not often hazardous for humans, but is harmful to plumbing because it deposits out again, eventually clogging the flow through the pipes. Hardness is not to be confused with acidity—they are two separate problems.

Hard water not only causes deposition in pipes, but is particularly problematic in heat exchangers, such as hot water tanks or hydronic systems. The deposits choke off the flow, or insulate the pipe so that the heat exchange is much reduced. A piece of metal is often inserted in the water tank to divert the deposition rather than allowing it to form on the heat exchanger. This is called an *anode.*

Hardness also interferes with the cleaning and lathering capabilities of detergents and soaps. Since there are already ions dissolved in the

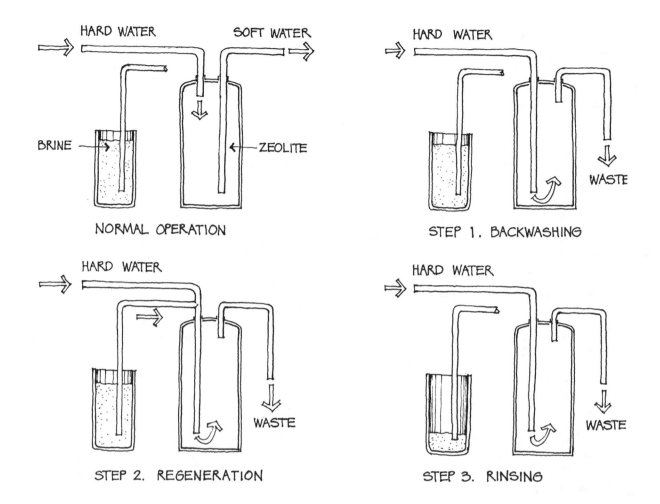

NORMAL OPERATION

STEP 1. BACKWASHING

STEP 2. REGENERATION

STEP 3. RINSING

WATER SOFTENER RECHARGE

water (from the minerals), getting additional dirt to dissolve becomes more difficult. Sometimes the minerals actually coagulate with the soap to form a soft paste in the plumbing and a fine film or build-up in whatever is being washed.

Water may be *softened* by removing the mineral ions, or by combining them with something that will not solidify when the water is heated. This is done using the *zeolite* or *ion exchange* process. Such a water softener consists of two tanks. The first contains the zeolite mineral and the second contains salt crystals. The water to be softened is passed through the zeolite tank.

Eventually the zeolite needs to be recharged. It is first backwashed to free clogged material, which is simply drained away. It is then regenerated with brine from the salt crystal tank. The sodium ions from the salt exchange with the magnesium and calcium ions which were trapped by the zeolite, and this is also drained away. Finally, the zeolite is rinsed to remove excess salt, and then returned to normal use.

Carcinogens

Unfortunately, natural minerals are not the only contaminants found in *ground water*. Many other substances have been added to our streams and water supplies, or even to the ground around wells or water supplies, which eventually are passed into the ground water. Most notable are the *carcinogens* (cancer causing agents) such as *PCBs (poly chlorinated biphenyls), DDT* and other insecticides, and *asbestos* fibers. Some areas of the country have experienced major contamination of ground water from the improper disposal of dangerous chemicals.

Consequently, in recent years, our water supply, which we used to trust implicitly, has become more suspect, and as a result, the use of bottled water has increased substantially. But what of the major portions of society who cannot afford bottled water?

Disease

Another type of contamination relates not to dissolved chemicals, but rather to the presence of bacteria or viruses in the water supply. This comes, most often, from the improper disposal of human or animal waste or other organic materials that, in decay, provide the source and breeding ground for disease-causing bacteria and viruses.

The traditional treatment for public water supplies consists of settling out contaminants, sometimes by adding a coagulant such as alum, and letting sediment settle out. *Chlorine* may be added to kill or reduce bacteria. The level is usually about 0.5 ppm (parts per million). If the level exceeds 1 ppm, a distinct chlorine taste is noticeable. *Fluorine* is sometimes added as well, to improve resistance to tooth decay. Oxygen is usually present, but if not, the water is oxygenated by passing it through a spray or a waterfall. The oxygen makes it more fit for human consumption, but also increases the rate at which ferrous (iron based) fittings *rust* or *oxidize*.

WATER PRESSURE

As mentioned earlier, water is usually supplied under pressure, which ordinarily should be sufficient to provide a reasonably high flow rate to the upper stories of a building. However, this is not always the case; water is fairly heavy, and lifting it requires a great deal of pressure.

Lift

To understand the relationship between pressure in pounds per square inch (psi) and static head (inches or feet of water that could be supported by a given pressure), one should know that a pressure of one psi is capable of lifting a column of water 2.3 feet high. Thus 10 psi could lift water 23 feet and 100 psi could lift it 230 feet. It is often necessary to have some pressure left at the top of the lift to flush a toilet or other fixture, as shown in the following example.

Example #1

A 10-story building has a floor-to-floor height of 12 feet. A pressure of 15 psi is required to flush a toilet. What is the required water pressure at the base of the building?

Solution:

Total lift = 10 stories × 12 feet/story
           = 120 feet

Since 2.3 feet is equivalent to 1 psi, 120 feet is equivalent to 120/2.3 = 52.2 psi.

Required water pressure
= 52.2 psi for lift + 15 psi for flush
= *67.2 psi*

If the pressure at the street level, say the street water main, is less than 67.2 psi, other means must be employed to supply the toilets at the upper floors. These will be discussed shortly.

Note that when converting from pressure (psi) to lift (feet) we use 2.3 feet per psi. When converting from height or lift (feet) to pressure (psi), we use 1/2.3, or .433 psi per foot.

Different fixture types require different pressures, which vary from 7 or 8 psi min for a faucet to 30 psi min for a hose bibb. Very high pressures cause undue wear on the valve seats and washers. Therefore, if pressures exceed 80 psi, a pressure regulator should be installed, which keeps the pressures in the 40 to 60 psi range.

As mentioned in Example #1, if the pressure in the water main is insufficient to supply the upper stories of a building, special systems must be employed. These include the downfeed system, the pneumatic tank system, and the tankless system.

In the *downfeed* system, a tank mounted on the roof supplies water to the upper stories. The tank is filled to a certain level by water from the main that is boosted by a pump in the basement. (*Note*: Water can be pushed up from below to just about any height. However, it can never be sucked up from above to any height greater than 33 feet, which is the static head equivalent of atmospheric pressure of 14.7 psi.) Since the roof tank supplies the upper floors, the pressure is determined by the height of the tank above a given floor, and not by the pump. Thus, the pressure at any level is constant. The disadvantage of the downfeed system is the considerable weight added to the roof, which results in a heavier and more expensive structure, particularly in earthquake-prone areas.

To avoid that, the *pneumatic tank* system uses a pressurized tank in the basement to supply higher levels. Some air is left in the tank, which can be compressed to act as a spring pushing up on the water in the tank (water does not compress). This takes up some floor space in the basement, and causes some air to be dissolved into the water.

The *tankless* system requires one or more variable speed pumps which constantly turn on or off or run at varying speeds to provide sufficient pressure at whatever demand rate the building requires. This tends to wear out the pumps more rapidly, but requires hardly any floor space and no additional roof space or structure.

Friction

The remaining consideration is the *flow rate*, and the resulting *pressure losses* due to *friction*. This is similar to the friction losses inside HVAC ducts as discussed in Lesson Five. The friction loss is a function of the diameter of the pipe and the flow rate through it. The smaller the diameter, the greater the friction at a constant flow rate. The greater the flow rate, the greater the friction at a given diameter. Valves, tanks, and other devices in the line also add friction. Thus, the vertical pipes in the building and the pipe from the street to the building represent additional losses (at design flow rate) that must be considered in addition to the lift requirements. The water meter itself represents a pressure loss depending on flow rate, from 1 to 25 psi for flow rates from 4 to 1,000 gpm.

Example #2

If there is an additional pressure drop of 12 psi due to friction in Example #1, what is the required water pressure at the base?

Solution:

$$P_{tot} = 52.2 \text{ psi (for lift)}$$
$$+ 15 \text{ psi (for flush)}$$
$$+ 12 \text{ psi (to overcome friction losses)}$$
$$= 79.2 \text{ psi (min)}$$

Hot Water Systems

There are usually two complete systems in a building, one which supplies cold water directly to the fixtures, while the other supplies cold water to a storage tank that is heated (the water heater), and then supplies the hot water to the faucets.

The water heater is always pressurized, and is rated in terms of its *volume* and its *recharge* rate. The volume is the capacity of the tank in gallons, and the recharge rate is the length of time that the tank will take to reheat itself after it has emptied out its volume of hot water.

There are several other variations. One is the solar hot water heater discussed in Lesson Four.

Another variation is the continuous loop system. Rather than allowing the water in the hot water pipes to cool off, thus requiring the user to waste all the water between the faucet and the heater before obtaining hot water, the hot water is continually pumped around a closed loop in the building.

There is a small, steady heat loss from the hot pipes, but no wasted water, since the user gets hot water almost instantaneously at the tap. In any case, it is important that all hot water pipes be insulated, since this is always the most cost effective first step, to conserve energy.

A totally different approach is the *in-flow* heater or *instantaneous* heater, in which only cold water is supplied to each fixture. The cold water is heated when the hot water faucet is turned on, or just prior to the expected need. This may be done automatically or manually. It is, of course, much more efficient, since no heat is wasted and no hot water is lost. The first cost is usually higher, and the convenience (flow rate, etc.) is often not as great, but such systems hold real promise. They may consist of electric resistance coils or small gas burners or heat exchangers.

Thermal Expansion

As with other building components, pipes expand and contract due to changing temperatures. This doesn't affect diameter very much, but it affects length a great deal. The change in the length of a pipe may be expressed by the formula:

$$\Delta L = Lk(T_2 - T_1)$$

where
   ΔL = the change in length
   L = length
   k = coefficient of expansion
   T_1 = original temperature
   T_2 = final temperature

TABLE 2.1 – THERMAL EXPANSION COEFFICIENTS	
Material	**Coefficient**
Steel	6.5×10^{-6}
Cast Iron	5.6×10^{-6}
Copper	9.8×10^{-6}
PVC	35×10^{-6}

Example #3

The temperature of a 100 ft. length of copper pipe increases from 65°F to 160°F when hot water starts to run through it. How much does it expand?

Solution:

$$\Delta L = Lk(T_2 - T_1)$$
$$= 100 \text{ ft. } (9.8 \times 10^{-6}) (160° - 65°)\text{F}$$
$$= 0.0931 \text{ feet}$$
$$= 0.0931 \times 12 = 1.12 \text{ inches}$$

That is quite a bit of movement and may present problems, especially if the pipe is fastened to a wall.

Therefore, pipe supports on long runs should be flexible, especially for hot water pipes. Some pipe supports actually have rollers, so that the pipe can move horizontally while it is supported vertically. Pipes should be supported every 4 feet for plastic, 6 feet for copper and 12 feet for steel.

WASTE SYSTEMS

While the major consideration with water supply is to keep it uncontaminated, the major consideration with drainage is the opposite, that is, to keep it from causing contamination. For this reason, and for the benefit of everyone downstream, *sanitary waste* from inside buildings is usually kept separate from *storm drainage*, which is the runoff of rain water from outside buildings.

Sanitary Waste

Sanitary waste water is always assumed to be contaminated, because sometimes it is. This is because the organic matter in sanitary waste is in various stages of decay. This decay produces by-products that are dangerous to health, obnoxious in odor, and sometimes flammable, such as methane gas, and laden with bacteria that are likely to cause disease.

The trap under a sink, for example, is not just to catch your contact lenses before they disappear forever. The trap remains full of water even after the remainder of the waste has been transported away by the water flow. The water in the trap prevents the methane or sewer gas created by the decomposition of the sewage from passing back up the drain pipe into the occupied spaces in the buildings.

There are two categories of sanitary lines: *soil lines*, which carry the water from toilets, urinals, and similar fixtures, and *waste lines*, which carry all other waste water from inside the building.

To relieve the pressure that builds up, and also to break the suction or siphoning action that occurs when water passes down through the system, sanitary drainage is connected to vents which rise out of the building to the open air above. There are three types of venting: vent stacks, stack vents, and soil stacks. The *soil stack* is a large pipe into which all of the soil and waste lines from one or more levels empty. It is open to the outside air at the top. The *vent stack* is a smaller pipe that is the air intake line for all the fixtures, and that is also, separately, open to the outside air at the top. In the soil stack, the section *above* the highest fixture is called the *stack vent* (as opposed to vent stack) and vents the soil stack. The vent stack is a "stack" of vents, the stack vent is something that vents the top of the soil stack. If this is confusing, we suggest that you look at the illustration on the following page. The minimum diameter for a vent is 1 1/4" or half the diameter of the drain it services, whichever is larger.

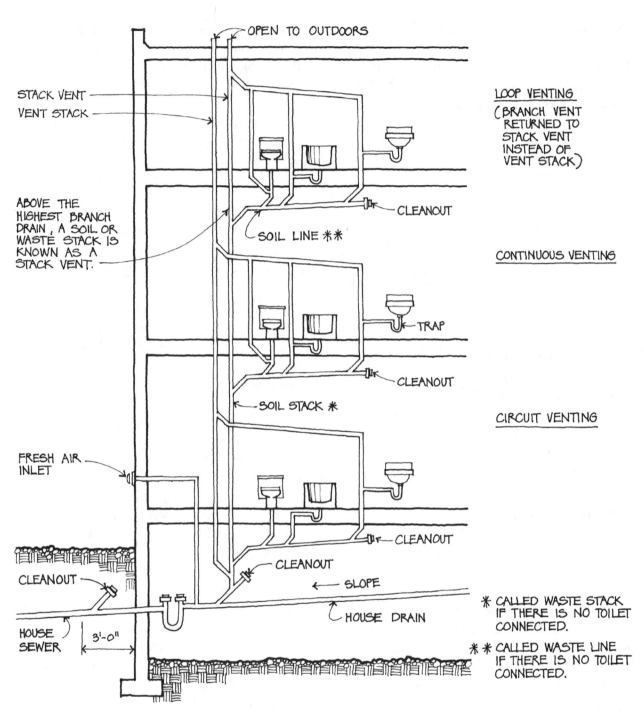

OPEN TO OUTDOORS

STACK VENT

VENT STACK

LOOP VENTING
(BRANCH VENT RETURNED TO STACK VENT INSTEAD OF VENT STACK)

ABOVE THE HIGHEST BRANCH DRAIN, A SOIL OR WASTE STACK IS KNOWN AS A STACK VENT.

CLEANOUT

SOIL LINE ✱✱

CONTINUOUS VENTING

TRAP

CLEANOUT

SOIL STACK ✱

CIRCUIT VENTING

FRESH AIR INLET

CLEANOUT

CLEANOUT

CLEANOUT

SLOPE

HOUSE DRAIN

CLEANOUT

HOUSE SEWER

3'-0"

✱ CALLED WASTE STACK IF THERE IS NO TOILET CONNECTED.

✱✱ CALLED WASTE LINE IF THERE IS NO TOILET CONNECTED.

SANITARY STACKS

Typical Layouts

The material used most often for sanitary lines is cast iron pipe, although ceramic pipe is sometimes used for lines outside the building, and copper or galvanized steel may be used for

vents. Plastic pipe may also be used, but is usually limited to residential applications.

The average toilet uses from three to five gallons of water per flush. Needless to say, that is quite a bit of clean water. This is true of *reservoir* or *tank type* toilets, which drain a tank through the waste bowl and then refill the tank, and of *flush valve* or *flushometer* toilets, which turn on water at a high rate of speed for a short time. In order to conserve water, several alternatives have been proposed, especially for residential applications where low volume is expected. The simplest alternative is the use of a smaller reservoir, which drops the water from a greater height, giving it a greater velocity.

More innovative is the *composting toilet*, in which no water at all is used, and the waste is stored below and vented to avoid noxious or unhealthy odors from collecting in the building. The biodegradable garbage from the kitchen also drops directly into the tank and eventually the combination provides a rich natural fertilizer. One example is the Clivus Multrum, a brand name for a Swedish company that manufactures composting toilets, which has a recycling time of approximately two years. After the toilet has been in use long enough, there is a steady supply of fertilizer.

Another, more complicated, concept separates urinal and toilet lines (soil lines) from sink and shower lines (called *grey water*). This does not reduce the amount of water per flush, but by keeping the organic wastes separate from the wash water, it allows the wash water to be processed and recycled on site, or at a small community scale. Although certain detergents must be avoided, this concept can be ecologically beneficial since it reduces the amount of potable water that is wasted.

HANDICAPPED ACCESS

Introduction

Although the standards for barrier-free architecture vary somewhat from state to state, the exam is based on the national standard, which is ANSI A117.1. Candidates should therefore become familiar with the handicapped provisions contained in that publication.

Rest Rooms

Toilet stalls for the handicapped in public rest rooms present a number of problems. To enable a person in a wheelchair to leave the stall, the door must swing outward, and he/she has to be able to open it. This usually means turning around, which requires a clear circular space of 5'-0" in diameter at a height of 10" above the floor. This is ideal and is in excess of the minimum ANSI requirements. Although stalls narrower than 5'-0" are permitted by the ANSI Standard, they present problems to a handicapped individual, who must be able to open a door behind his/her back.

In addition, once inside the stall, the individual must be able to transfer to the seat. It is best if the seat is at the wheelchair height of 1'-7", so that the return transfer is not to be "uphill." In stalls that are less than 5'-0", there must be grab bars on each side wall for the individual to lift himself or herself and turn around. Again, the 5'-0" stall is preferable, because it allows a side, or parallel transfer, where the wheelchair-bound individual backs up next to the seat and shifts across onto it. This also requires grab bars, one on the side wall and one on the rear wall. In both cases the grab bars should be set at a height of 2'-9" to 3'-0".

Lavatories

There should be at least one lavatory that has sufficient clearance underneath so that a wheelchair can be moved up under it. The fixtures should have large lever controls rather than a small handle, or most absurd, a single, little round ball. Exposed hot water lines under such sinks should be well insulated. The mirror above this sink should be tilted slightly forward, so that a seated individual can easily see himself or herself. It is also a good idea to place the faucets on the side of the lavatory instead of at the back, if the counter is deep.

Drinking Fountains

Two different drinking fountains are often used, one at a height of 36" to 39" for adults without physical impairment, and the other at 32" (preferred) to 36" (maximum) with a clear space below for wheelchair access. The lower fountain should protrude from the wall as far as possible, consistent with safe traffic flow. Some paraplegics dehydrate rapidly, and access to a drinking fountain is therefore important.

Baths and Showers

Handicapped people must be able to bathe or shower in hotel rooms, and there are many hotel rooms in which this is simply not possible. Bath tubs should be supplied with grab bars, and at least one tub in every hotel should be available with a seat, roughly at wheelchair height, or at least at the height of the tub edge. Elevating the entire tub slightly to achieve this is an excellent idea. At least one shower should have a minimum height curb, or no curb at all, and a door wide enough (33") to allow a wheelchair to roll in. Again, a seat at wheelchair height is preferable, as well as a flexible hose and nozzle arrangement for the shower head. If space permits, a shower stall with a clear space of 5'-0" in diameter is best. If nec-

essary, the seat may intrude to create a 4'-0" circle, as long as 5'-0" is available at a 10" height above floor level, to clear the feet and the front wheels of the chair. Again, these dimensions are ideal and in excess of the minimum ANSI requirements.

In all of the above situations, common sense and a bit of thought are the primary considerations. Simply consider how a wheelchair-bound individual would go about using the facilities, and design accordingly.

MAINTENANCE

Unfortunately, every sanitary system eventually clogs up. However, there are ways of delaying that event, and there are fittings made to deal with the problem when it does occur.

Interceptors

Interceptors are designed to catch grease, hair, oil, string, rags, money, toothbrushes, etc., which get into the system. They are required by code for certain establishments such as restaurants,

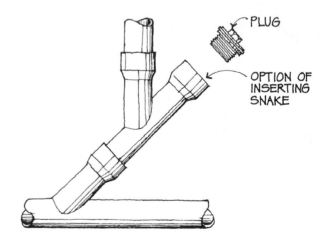

PLUG

OPTION OF INSERTING SNAKE

CLEANOUT

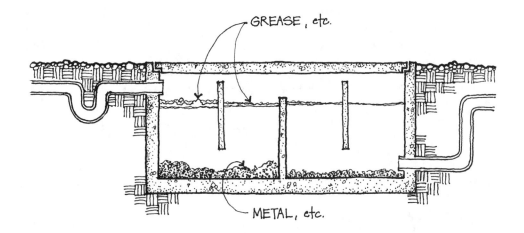

GREASE, etc.

METAL, etc.

INTERCEPTOR

which produce sufficient grease to create problems for the sewage system and treatment plant. Every interceptor is provided with some means of cleanout access, so that trapped material may be removed and discarded. Please note that a "trap" is the U-shaped line which keeps methane from the sewer system out of the building, while an "interceptor" is a device which keeps grease and other waste matter from the building out of the sewer system.

In addition to the cleanout for each interceptor, there are *cleanouts* for the system itself. A cleanout is just a Y-shaped segment of pipe which serves an area otherwise difficult to access, where one arm of the Y has a plug screwed in at its end. At least one cleanout is required where the building drain joins the sewer system. Cleanouts should be placed every 50 feet in pipes under 4 inches in diameter, and every 100 feet in larger pipes. There should also be a cleanout at every corner where the pipe changes direction more than 45°. When the system does clog up, the plug is unscrewed and a "snake," a length of flexible cable with a bit at the end, is inserted and rotated until the clog

is broken up. The clog may be removed through the cleanout, or flushed down the sewer, or both.

Manholes

Manholes are the equivalent of cleanouts for large lines, ten inches in diameter and greater. They occur at 150 foot intervals and wherever a new line is joined to an existing one. They also provide access for inspection.

SEWAGE TREATMENT SYSTEMS

Public Systems

In public sewage systems, all sewage is treated in a waste treatment plant before being returned to the nearest body of water. This is done by directing the flow into settling chambers that allow the solids to settle out. The remaining liquid sewage is treated using activated sludge, which is simply a rich mixture containing the necessary bacteria to digest the waste materials. The resulting clear water is chlorinated and returned to the surrounding water. The solid waste is put in an anaerobic digester (no oxygen present) and reduced in volume and digested by different bacteria. The result, *sludge*, is dried and either put in land fill, or occasionally used as fertilizer. Unfortunately, because of pollution today, sludge is usually contaminated, which precludes its use for most farming and residential applications.

When a public sewage system is unavailable, some form of private sewage treatment is necessary, such as a septic tank with leach field or a cesspool.

Cesspools

Cesspools provide the cheapest sewage treatment, but also the least desirable, and have been outlawed by code almost everywhere. A cesspool is simply an underground chamber with a porous bottom and porous walls. The sewage soaks into the surrounding ground until everything gets clogged up. This means that eventually a new cesspool must be dug and the sewage lines rerouted to it.

Septic Tanks and Leach Fields

A *septic tank* is a lined chamber, or often a steel tank, into which the sewage collects. The solid material deposits out, and the liquid waste passes on into a leach field. The solid material must be removed every few years. Septic tanks are sized based on an expected flow of 100 gallons per day per person, with a minimum capacity of 500 gallons.

The *leach field* or *tile drain field* is simply a grid of ceramic pipe laid underground, not quite together end to end, so that the liquid leaks out, over a bed of gravel, which filters the liquid waste before it seeps into the soil. Where the soil is impermeable, a large basin is dug and filled with sand. The liquid waste leaks into the sand, and after it is filtered by the sand, it is collected at the bottom of the basin, chlorinated, and returned to a local body of water.

STORM DRAINAGE

Surface runoff from rainfall is generally kept separate from sanitary waste because it is basically clean, and does not need to be treated in the same way as the more hazardous wastes. If it were piped into the sanitary system, it would overload the system. This sometimes happens with older systems or when rain water is illegally drained into the sanitary system. At such times, sewage treatment plants allow the raw sewage to flow along with the rain water out into the open stream or ocean discharge.

Water Table and Ground Water Recharge

Before urbanization covered much of the land with buildings, parking lots, and streets, most of the rain water soaked into the ground before running off. Thus, there was a steady flow of *ground water* to *recharge* the *water table* (underground water level). It also meant that the likelihood of a local flood was lower than today. The dual problems of insufficient water retention for wells and springs and excessive runoff and flooding can be ameliorated by intelligent handling of storm drainage at the building and site.

Many suburban corporate campuses now have large depressed recreational areas, which are allowed to flood during heavy rainfall. Much of the water in the flooded area soaks into the ground, while the flood crest passes. These areas must still be properly drained, to avoid breeding mosquitoes, but at a limited rate. This drainage is done by means of *swales* and *catch basins*.

Swales

Swales are shallow V-shaped sloping channels in the grass that take the surface runoff to points where it may be collected and/or disposed of.

Catch Basins

Catch basins are similar to manholes, except that the top has a grate instead of a cover. They are placed at the lowest point in a swale or a depression, to collect the runoff and pass it into the storm drainage system, which empties into the local stream or lake.

MATERIALS AND METHODS

Historically, a number of different materials have been used for plumbing. At one time, lead was actually used for pipe, primarily for drainage, because it was malleable (bendable by hand) and easy to work with. Lead pipe is now found only in old buildings. Each plumbing material has its own characteristics and typical connections.

Steel

Untreated steel was originally called black iron, because of its color. It was very susceptible to rust and corrosion and was replaced by *galvanized* steel pipe, steel pipe with a thin layer of zinc bonded to the surface to make it comparatively rust-resistant. Wall thicknesses for various pipe diameters are standardized by schedules, of which *Schedule 40* is the most common.

Steel pipe is typically joined mechanically. The two ends to be joined are threaded with a sloping thread, joint compound or tape is applied to seal the minute cracks or gaps, and the two pieces are screwed into a connecting collar with internal threads. When used in drainage systems, the

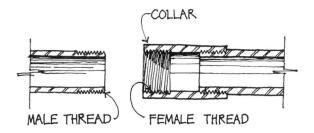

STEEL PIPE CONNECTIONS

two ends are often clamped together with a rubber sleeve, a steel jacket, and two steel band clamps.

Copper

Copper tubing is often used for supply piping, and is considered to be the best material for that purpose. Copper does not rust, and is resistant to corrosion because oxidation builds up a thin film that protects the copper. The wall thicknesses are much less in copper than in other materials. There are three categories of copper tubing: Type K, Type L, and Type M, with M being the most common and having the thinnest walls.

Copper pipes are joined by a form of *soldering* called *sweating. Flux* is applied to clean the

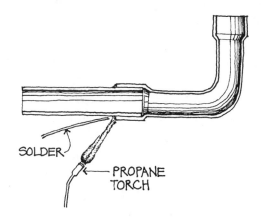

COPPER PIPE CONNECTIONS

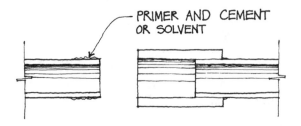

PLASTIC PIPE CONNECTIONS

surfaces to be bonded, and the sections are heated to a temperature that will melt the flux. The joint is assembled, using a sleeve or elbow that fits closely over the two sections. The solder is applied to the joint, melts, and is drawn into the joint by the capillary attraction between the two surfaces. When the joint cools, the pipes are structurally joined and completely sealed. One advantage of this method is that it is reversible. If the joint is reheated, the solder will melt, allowing the pipes to be slid apart.

Plastic

The prime competitor of copper is plastic pipe. Plastic pipe is available in two types, the PVC pipe, usually used for supply piping (white, with light blue lettering) and the ABS pipe, which is typically used for drainage (larger, black, with white lettering). Both are joined in the same manner, but using different solvents or cements. Plastic pipe does not corrode, and does not allow the electrolysis that deposits mineral ions. This makes it very durable when properly used. It deteriorates, however, when exposed to ultraviolet light, which is present in direct sunlight. For this reason, plastic pipe should never be used in exposed locations above ground or outside walls. When installing irrigation pipe, for example, all above ground fixtures should be steel or a different plastic or vinyl compound, which is typically green in color.

Plastic connections are made in a manner similar to copper connections. The surfaces are primed, and after the primer has dried, a cement or solvent is applied and the joint is slid together. The connection is structurally complete and sealed. However, the process is not reversible. There is no way to pull the pieces apart if they have been properly joined. Therefore, if changes are necessary, the original connection must be cut out with a saw (not very difficult) and new pipe provided for the gap created.

VALVES AND FIXTURES

Valves

There are several kinds of valves in a plumbing system. A *gate valve* is intended to be entirely on or entirely off. It has a minimum restriction when fully open, but causes a great deal of turbulence when partially open.

Globe valves are used not only to turn water on and off but also to meter or throttle the flow at intermediate rates. Globe valves restrict flow even when wide open.

A *check valve* is essentially a *backflow preventer*. Its purpose is to prevent water from moving backwards through the system. This is critical

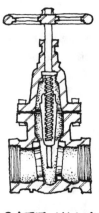

GATE VALVE

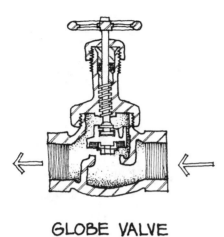

GLOBE VALVE

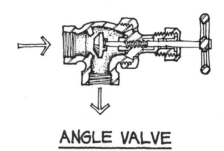

ANGLE VALVE

in avoiding contamination of the community supply, for example, to prevent backflow from an irrigation system. The simplest version consists of a flap that opens in the direction of the desired flow, but is pressed shut by flow in the other direction.

A preferable type is a ball that is spring loaded, so that it is pushed away from the mouth of the pipe by a desired flow, pops back with still water, and is pressed tightly shut by any reverse flow.

Typical plumbing fixtures include lavatories, sinks, showers, and appliance hookups such as

dishwashers and automatic icemakers. Such fixtures include a valve with a metering or flow restriction capability. These used to be similar to globe valves and were called *angle valves* or *screw* and *seat* or *washer* and *seat* valves. They screwed a washer down against a seat to shut the flow off, or opened and regulated the flow by screwing progressively away from the seat. These were twist handle faucets and hose bibbs. Now there is a myriad of patented single handle systems and cartridge systems that shut off more completely and easily when new, but that tend to cost a great deal when they need maintenance, and become automatically obsolete when the manufacturer "improves" the model. A new washer costs 70 cents, but a new cartridge costs $24.00, and a 15-year-old cartridge is often impossible to replace.

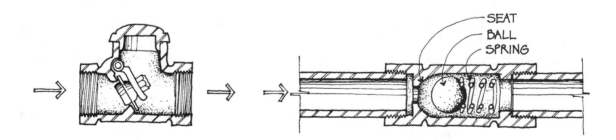

CHECK VALVES

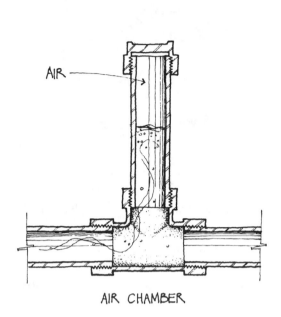

AIR CHAMBER

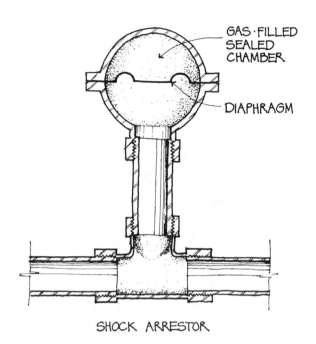

GAS·FILLED
SEALED
CHAMBER

DIAPHRAGM

SHOCK ARRESTOR

SURGE ARRESTORS

Pressure release valves are safety devices that keep systems from exploding by opening when the pressure exceeds some preset maximum. They are usually placed over a drain or in some area where the released water or steam will do no damage. They are required on water heaters, for example.

Surge Arrestors

One common source of annoyance is *water hammer*, that thumping or rattling sound that occurs when a faucet is shut off rapidly. This occurs upstream of the fixture where a long straight pipe run ends in a corner. The water that was flowing in the pipe is suddenly halted, and the inertia hammers the corner at the end of the run. In extreme cases it can actually burst out the end of the pipe. The solution is to insert a cushion or device with a damping effect somewhere in the system. These devices are called *shock arrestors* or *surge arrestors*. Alternatively, a length of ver-

tical pipe with nothing but air in it may be included in the system. The air compresses and bounces the water back at the original pressure, absorbing the shock.

Fixtures and Flow Rates

Different fixtures have different flow requirements. In order to determine the required size of pipe, an arbitrary unit is used for pipe sizing that takes into account the likelihood that all the fixtures will be in use at the same time. This is called the *fixture unit (FU)*. The relationship between gallons per minute (gpm) and FU is not constant, but varies with the number of fixture units. For example, 1,000 FU is equivalent to 220 gpm, yet 2,000 FU is not double that, but is only 1.5 times as much, or 330 gpm.

Two tables are necessary to determine pipe sizes: FUs per fixture type, and pipe sizes for total FUs.

TABLE 2.2 – FIXTURE UNITS PER FIXTURE TYPE		
Fixture	**FU Value**	**Minimum Trap Size, Inches**
Bath tub	2	1 1/2
Drinking fountain	1/2	1 1/4
Lavatory	1	1 1/4
Shower	2	1 1/2
Kitchen sink	2	1 1/2

Reproduced by permission of the International Conference of Building Officials.

Another important consideration is that waste water must not be allowed to contaminate the supply water. For example, faucets should never extend down into the sink where standing water might cover the tip of the faucet. There should always be a two inch air gap designed into the system to prevent siphoning, by making the overflow on the sink or tub two inches lower than the faucet nozzle. This avoids the waste

water being sucked into the faucet if there is a loss of pressure in the supply line, which could occur, for example, when several faucets two stories below are turned on full, and the pressure temporarily drops or becomes negative on this level. Where siphoning is likely, a *vacuum breaker* must be installed, for example, over a flushometer valve that could back up if there were a loss in the supply line pressure.

SUMMARY

In this lesson we have discussed water supply, sanitary waste and storm drainage. We have identified some typical materials and connections, typical problems, and common systems, fixtures and valves.

Although exam questions may sometimes require calculations, more often candidates are tested on their understanding of basic concepts and their ability to recognize the various elements of water systems.

TABLE 2.3 – PIPE SIZES FOR FIXTURE UNIT TOTALS						
Pipe Size Inches	**Drain or Sewer**			**Horizontal Fixture Branch**	**Stack of 3 Branch Intervals or less in Height**	**Stacks Greater Than 3 Branch Intervals***	
	1/8" Fall	**1/4" Fall**	**1/2" Fall**			**Total for Stack**	**Total for 1 Branch Interval**
1 1/4		1	1	0	0	0	10
1 1/2		3	3	3	4	8	2
2		21	26	6	10	24	6
1 1/2		24	31	12	20	42	9
3	36	42	50	20	48	72	20
4	180	216	250	160	240	500	90
5	390	480	575	360	540	1,100	200

*Branch Interval—A distance along a soil or waste stack corresponding in general to a story height, but not less than 8 feet, within which the horizontal branches from one floor or story of a structure are connected to the stack.

Reproduced by permission of the International Conference of Building Officials.

LESSON 2 QUIZ

1. Which of the following statements is correct?

   I. Rain water is slightly acidic.

   II. Rain water is slightly alkaline.

   III. Rain water tends to corrode metal pipes.

   IV. Rain water is often less healthful than ground water.

   A. I and III

   B. II only

   C. I, III, and IV

   D. II, III, and IV

2. Which of the following statements is correct?

   A. Ground water is naturally soft.

   B. Hard water is often hazardous to humans.

   C. Ground water is naturally hard.

   D. Hard water can be controlled by the use of acid neutralizers.

3. Which of the following statements is correct?

   A. Acidic water causes deposition, so pipes should be oversized.

   B. Acidic water can be controlled by using the zeolite or ion exchange process.

   C. PVC pipe is appropriate for supply piping in exposed locations above ground.

   D. Copper and PVC are resistant to corrosion and therefore appropriate where water is acidic.

4. Which of the following represents an extremely acidic pH reading?

   A. pH = 5

   B. pH = 7

   C. pH = 9

   D. pH = 17

5. Which of the following piping materials is heated when two pieces are joined?

   A. Steel

   B. Copper

   C. Iron

   D. Plastic

6. If the street main provides 100 psi, and there is a loss of 10 psi due to friction between the street and the building, how high can we place a faucet if the required pressure at the fixture is 10 psi?

   A. 74 feet

   B. 98 feet

   C. 124 feet

   D. 184 feet

7. A downfeed system is different from a pneumatic system in that

   I. a downfeed system is necessary only when the street main pressure would not supply the upper floors.

   II. a downfeed system is not pressurized.

   III. a downfeed system has no storage tank.

   A. I only

   B. II only

   C. III only

   D. I and III

8. Why is it necessary to keep storm drainage separate from sanitary sewage?

   I. Storm drainage is more polluted than sanitary sewage.

   II. Storm drainage tends to flow at high volumes, which are not treatable by the sewage plant.

   III. Storm drainage is not pressurized but sanitary drainage is.

   A. I only

   B. II only

   C. III only

   D. II and III

9. The purpose of a sanitary trap is to

    A. catch valuable objects before they enter the sewage system.

    B. catch dangerous chemicals before they enter the sewer system.

    C. keep storm drainage and sanitary sewage separate.

    D. keep sewer gas from backing up into the building.

10. What is a catch basin?

    A. An inlet into the drainage system for surface runoff

    B. A device which keeps grease and other waste matter out of the sewer system

    C. A type of cleanout which provides access for inspection and maintenance

    D. A lined chamber into which sewage collects

BASIC THERMAL PROCESSES

INTRODUCTION

If we look at the history of shelter, we find that the original concerns were primarily related to thermal comfort. In fact, the word "shelter" may be defined as protection from the elements. Architecture has addressed many other issues since then, but the original considerations still remain. In this lesson, we will discuss some of the basic concepts and formulas related to thermal comfort.

The thermal processes of a building are analogous to those of the human body, in which the skin keeps the heat in or out and the circulation systems regulate the temperature. The skin of a building and that of the body perform the same function. The circulation system of a building consists of the mechanical equipment. The form and exposure of the building's skin is critical, and is determined by the architect. The circulation system is determined by the mechanical engineer, and as long as the architect provides sufficient space and doesn't create any insoluble problems the building can be kept appropriately warm or cool. To understand form, skin and exposure, and to avoid creating insoluble problems, we will first address basic heat transfer processes.

BASIC PHYSICS OF HEAT TRANSFER

There is a difference between heat and temperature—they are related but are not the same. Temperature is a measure of stored heat energy, but temperature is never transferred, only heat energy is.

Furthermore, heat is sometimes transferred without a change in temperature (such as the melting of an ice cube). When transferred heat causes a change in temperature it is called *sensible heat.* When it causes a change of state (ice to water or water to vapor) it is called *latent heat.* Heat always flows from a hotter object to a cooler object.

Two objects at the same temperature may store different amounts of heat (concrete stores a lot more heat than carpet, for example). This storage capacity is called the *specific heat (C_p)* of a material, which is measured by comparing it to the storage capacity of water. A *British Thermal Unit (Btu)* is defined as the amount of heat energy required to raise one pound of water by 1°F. Thus specific heat is measured in terms of the number of Btu's required to change the temperature of a specific material by 1°F.

Radiation

Radiation is the method by which heat is transferred between two objects not in contact and not shielded from each other. When you stand in the sun, you are experiencing heat transfer from the sun by radiation, in spite of the fact that you are not touching it (conduction), nor are you above it (convection). Radiation is always taking place, but usually at a slow rate. All objects radiate at each other, even two people standing near each other. The *wavelength* of the radiation is based on the *temperature* of the object. Warm things radiate infrared, while really hot objects (i.e., red hot steel) begin to glow in the visible spectrum. If they get even hotter, they glow orange, and then white hot. The *rate* of radiative exchange is based on the surface *temperature* of the objects, the *viewed angle* and a property called *emissivity.* Surfaces

TABLE 3.1 – SPECIFIC HEAT (C_p)

Description	Density lb/ft³	Conductivity[b] λ Btu•in./ h•ft²•F	Conductance (C) Btu/h• ft²•F	Resistance [c] (R) Per inch thickness (1/λ) h•ft²• F/Btu	Resistance [c] (R) For thickness listed (1/C) h•ft²• F/Btu	Specific Heat, Btu/lb • deg F
MASONRY UNITS						
Brick, common[i]	120	5.0	—	0.20	—	0.19
Brick, face[i]	130	9.0	—	0.11	—	
Clay tile, hollow:						
1 cell deep.........................3 in.	—	—	1.25	—	0.80	0.21
1 cell deep.........................4 in.	—	—	0.90	—	1.11	
2 cells deep........................6 in.	—	—	0.66	—	1.52	
2 cells deep........................8 in.	—	—	0.54	—	1.85	
2 cells deep.......................10 in.	—	—	0.45	—	2.22	
3 cells deep.......................12 in.	—	—	0.40	—	2.50	
Concrete blocks, three oval core:						
Sand and gravel aggregate............4 in.	—	—	1.40	—	0.71	0.22
....................8 in.	—	—	0.90	—	1.11	
....................12 in.	—	—	0.78	—	1.28	
Cinder aggregate.....................3 in.	—	—	1.16	—	0.86	0.21
....................4 in.	—	—	0.90	—	1.11	
....................8 in.	—	—	0.58	—	1.72	
....................12 in.	—	—	0.53	—	1.89	
Lightweight aggregate.................3 in.	—	—	0.79	—	1.27	0.21
(expanded shale, clay, slate..........4 in.	—	—	0.67	—	1.50	
or slag; pumice):....................8 in.	—	—	0.50	—	2.00	
....................12 in.	—	—	0.44	—	2.27	

Note: Unless otherwise noted, all tables in this lesson are from the 1989 ASHRAE Handbook of Fundamentals and are reprinted by permission.

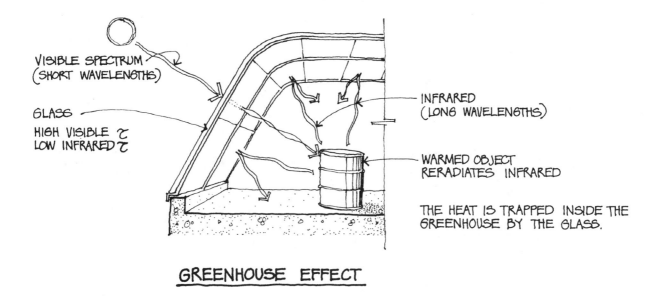

VISIBLE SPECTRUM
(SHORT WAVELENGTHS)

GLASS
HIGH VISIBLE τ
LOW INFRARED τ

INFRARED
(LONG WAVELENGTHS)

WARMED OBJECT
RERADIATES INFRARED

THE HEAT IS TRAPPED INSIDE THE
GREENHOUSE BY THE GLASS.

GREENHOUSE EFFECT

with high emissivities radiate at a rate higher than those with low emissivities.

The *emissivity* (ε) of a surface is a property of the material, and is usually the same as the *absorptivity* (α) at any *given wavelength.* The simplest example in the visible spectrum is color. Black surfaces have higher emissivities and absorptivities than white or shiny surfaces. Black surfaces heat up rapidly and cool off rapidly. Shiny surfaces heat up more slowly, but stay hot longer. The emissivity and absorptivity are often different between the infrared and visible spectrum. *Selective surfaces* are surfaces that have a high absorptivity in one wavelength (usually solar) and low emissivity in another (usually infrared). This means that the material stores incoming *solar* radiation without releasing it as *infrared* (i.e., a good solar collector panel).

The foil sometimes used on the back of fiberglass insulation has a very low emissivity, which reduces the radiative transfer across the air space inside the wall.

You are probably more familiar with *transmissivity* (τ), which is the measure of how easily a material allows radiant energy to pass through it. We normally consider glass to be transparent, since it has a very high transmissivity in the visible spectrum. In the infrared region, however, it has a very low transmissivity, which is what causes the "*greenhouse effect.*" Sunlight (short wavelength and visible portion of spectrum) is transmitted through the glass into a building, causing the materials inside to heat up. When they heat up, they reradiate, but in the infrared spectrum. This reradiated energy does not pass through the glass and is therefore trapped inside the building. This is similar to the behavior of a selective surface, only now the selection is in terms of what passes through, rather than what is absorbed.

One of the current questions society must examine is the greenhouse effect in the upper atmosphere. As more carbon dioxide is released into the air, the rate at which the earth reradiates into space changes. Many scientists believe that this is increasing the average temperature of the earth. This could be quite dangerous, because such climatic changes would cause melting in the polar caps, thus raising the levels of the oceans. At the same time, the snow line would

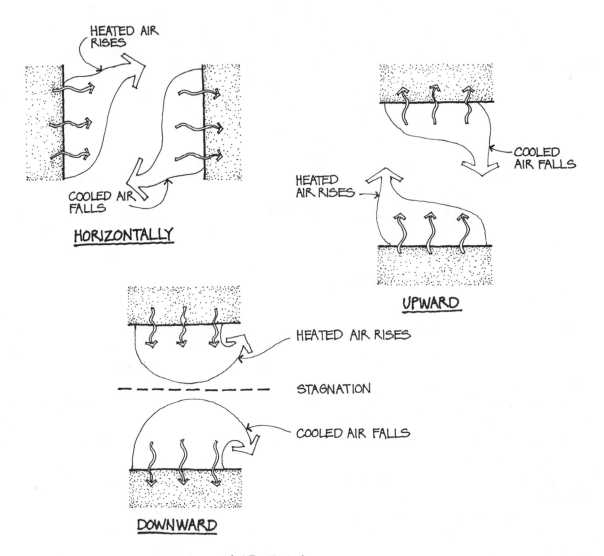

CONVECTION

rise in most mountains, reducing the amount of water stored for following summers. This would result in worse flooding in winter and worse droughts in summer. The question bears further examination, beyond the scope of this course.

If you are trying to warm a building, the greenhouse effect is a great benefit, but if you are trying to keep the building cool, it's a good reason to avoid horizontal skylights!

The *viewed angle* of a surface depends on the size of the surface and your distance from it. When you stand very close to a meat freezer it occupies a large angle of view, and you lose a surprising amount of heat to it. When you stand across the room, however, it occupies a smaller angle of view and there is consequently much less heat exchange. The average radiant temperature of your surroundings is called the *Mean Radiant Temperature (MRT)*, and is independent of air temperature. For example, when you are skiing on a sunny day, the air temperature

TABLE 3.2 — AIR SPACE RESISTANCES

Position of Air Space	Direction of Heat Flow	Mean Temp,[d] (F)	Temp Diff,[d] (deg F)	0.5-in. Air Space[c] Value of E[d,e]					0.75-in. Air Space[c] Value of E[d,e]				
				0.03	0.05	0.2	0.5	0.82	0.03	0.05	0.2	0.5	0.82
Horiz.	Up	90	10	2.13	2.03	1.51	0.99	0.73	2.34	2.22	1.61	1.04	0.75
		50	30	1.62	1.57	1.29	0.96	0.75	1.71	1.66	1.35	0.99	0.77
		50	10	2.13	2.05	1.60	1.11	0.84	2.30	2.21	1.70	1.16	0.87
		0	20	1.73	1.70	1.45	1.12	0.91	1.83	1.79	1.52	1.16	0.93
		0	10	2.10	2.04	1.70	1.27	1.00	2.23	2.16	1.78	1.31	1.02
		−50	20	1.69	1.66	1.49	1.23	1.04	1.77	1.74	1.55	1.27	1.07
		−50	10	2.04	2.00	1.75	1.40	1.16	2.16	2.11	1.84	1.46	1.20
45° Slope	Up	90	10	2.44	2.31	1.65	1.06	0.76	2.96	2.78	1.88	1.15	0.81
		50	30	2.06	1.98	1.56	1.10	0.83	1.99	1.92	1.52	1.08	0.82
		50	10	2.55	2.44	1.83	1.22	0.90	2.90	2.75	2.00	1.29	0.94
		0	20	2.20	2.14	1.76	1.30	1.02	2.13	2.07	1.72	1.28	1.00
		0	10	2.63	2.54	2.03	1.44	1.10	2.72	2.62	2.08	1.47	1.12
		−50	20	2.08	2.04	1.78	1.42	1.17	2.05	2.01	1.76	1.41	1.16
		−50	10	2.62	2.56	2.17	1.66	1.33	2.53	2.47	2.10	1.62	1.30
Vertical	Horiz.	90	10	2.47	2.34	1.67	1.06	(0.77)	3.50	3.24	2.08	1.22	0.84
		50	30	2.57	2.46	1.84	1.23	(0.90)	2.91	2.77	2.01	1.30	0.94
		50	10	2.66	2.54	1.88	1.24	0.91	3.70	3.46	2.35	1.43	1.01
		0	20	2.82	2.72	2.14	1.50	1.13	3.14	3.02	2.32	1.58	1.18
		0	10	2.93	2.82	2.20	1.53	1.15	3.77	3.59	2.64	1.73	1.26
		−50	20	2.90	2.82	2.35	1.76	1.39	2.90	2.83	2.36	1.77	1.39
		−50	10	3.20	3.10	2.54	1.87	1.46	3.72	3.60	2.87	2.04	1.56
45° Slope	Down	90	10	2.48	2.34	1.67	1.06	0.77	3.53	3.27	2.10	1.22	0.84
		50	30	2.64	2.52	1.87	1.24	0.91	3.43	3.23	2.24	1.39	0.99
		50	10	2.67	2.55	1.89	1.25	0.92	3.81	3.57	2.40	1.45	1.02
		0	20	2.91	2.80	2.19	1.52	1.15	3.75	3.57	2.63	1.72	1.26
		0	10	2.94	2.83	2.21	1.53	1.15	4.12	3.91	2.81	1.80	1.30
		−50	20	3.16	3.07	2.52	1.86	1.45	3.78	3.65	2.90	2.05	1.57
		−50	10	3.26	3.16	2.58	1.89	1.47	4.35	4.18	3.22	2.21	1.66
Horiz.	Down	90	10	2.48	2.34	1.67	1.06	0.77	3.55	3.29	2.10	1.22	0.85
		50	30	2.66	2.54	1.88	1.24	0.91	3.77	3.52	2.38	1.44	1.02
		50	10	2.67	2.55	1.89	1.25	0.92	3.84	3.59	2.41	1.45	1.02
		0	20	2.94	2.83	2.20	1.53	1.15	4.18	3.96	2.83	1.81	1.30
		0	10	2.96	2.85	2.22	1.53	1.16	4.25	4.02	2.87	1.82	1.31
		−50	20	3.25	3.15	2.58	1.89	1.47	4.60	4.41	3.36	2.28	1.69
		−50	10	3.28	3.18	2.60	1.90	1.47	4.71	4.51	3.42	2.30	1.71

Position of Surface	Direction of Heat Flow	Surface Emittance					
		Non-reflective ε = 0.90		Reflective ε = 0.20		Reflective ε = 0.05	
		h_i	R	h_i	R	h_i	R
STILL AIR							
Horizontal	Upward	1.63	0.61	0.91	1.10	0.76	1.32
Sloping—45 deg	Upward	1.60	0.62	0.88	1.14	0.73	1.37
Vertical	Horizontal	1.46	(0.68)	0.74	1.35	0.59	1.70
Sloping—45 deg	Downward	1.32	0.76	0.60	1.67	0.45	2.22
Horizontal	Downward	1.08	0.92	0.37	2.70	0.22	4.55
MOVING AIR (Any Position)		h_o	R	h_o	R	h_o	R
15-mph Wind (for winter)	Any	6.00	(0.17)				
7.5-mph Wind (for summer)	Any	4.00	0.25				

Surface	Reflectivity in Percent	Average Emittance ε	Effective Emittance E of Air Space One surface emittance ε; the other 0.90	Both surfaces emittances ε
Aluminum foil, bright	92 to 97	0.05	0.05	0.03
Aluminum sheet	80 to 95	0.12	0.12	0.06
Aluminum coated paper, polished	75 to 84	0.20	0.20	0.11
Steel, galvanized, bright	70 to 80	0.25	0.24	0.15
Aluminum paint	30 to 70	0.50	0.47	0.35
Building materials: wood, paper, masonry, nonmetallic paints	5 to 15	0.90	0.82	0.82
Regular glass	5 to 15	0.84	0.77	0.72

may be cool, but the radiant energy coming from the sun, and the reflection of the sun off the snow, combined with a little exercise, will make you quite warm. There is a special device called a *globe thermometer* that can be used to measure MRT.

Convection

Convection is the heat exchange process that occurs only in a fluid medium, such as air or a liquid. Hot air rising is an example of convection. Air expands when it is hot, which reduces its density and makes it lighter. The cool, heavier air falls, while the warm, lighter air rises.

Smoke rises in chimneys because it is warmer than the room air. The only material that expands when it gets colder is water, and only just before it freezes. That is fortunate indeed, otherwise ice would always form at the bottom of a lake rather than on the surface, and all the fish would die. (It would also be difficult to ice skate.)

Convection occurs in rooms all the time, especially in large atrium spaces, as well as inside wall cavities, such as between studs. Convection is the only means of heat transfer which is strictly directional. It never transfers heat *downward*. It can transfer heat *horizontally* by stirring the air, but not as rapidly as *upward*. When the top of a space is warmer than the bottom, and the hot air rises and stays there, it is called *stagnation*. The difference in pressure in a vertical space (positive or outward at the top and negative or inward at the bottom) is called the *stack effect*. The rising air tries to push out at the top, and it pulls air in behind it, down below. The stack effect can be significant in tall office towers, in which the elevator shafts can act like smokestacks.

Tables indicating the thermal resistances of air spaces have different values for the same thickness, depending on the orientation of the space (horizontal or vertical) and the direction of the heat flow (up or down). The orientation is more critical than the thickness.

The thin film of air which occurs next to a wall also provides a resistance, calculated as the inverse of the so-called *film coefficient (fi)*.

Conduction

Conduction is the heat transfer process that occurs when objects are in direct contact. Picking up a hot frying pan results in immediate (and painful) conduction. Conduction is not directional; there is no preference for up or down, only from hotter object to colder object. In buildings, conduction occurs inside a wall, transferring heat from inside to outside in cold climates primarily by the direct contact of the different layers in the wall. Each material has a different *conductivity (k)*, and *resistivity (r)*, which is the inverse of conductivity. Specific thicknesses of materials have calculated *conductances (C)* and *resistances (R)*. The resistance (R) is calculated from the thickness (x) and the conductivity (k), using the formula:

$$x/k = R$$

Insulation is often specified by the letter R followed by a number. For example, R - 19 insulation has a resistance of 19 ($ft^2°F$ hr / Btu). A complete wall assembly has a calculated "conductance," which represents all the interactions of the internal materials, including some radiation and convection, and which is called the *U value*. The U value is the reciprocal of the sum of the resistances (1 divided by the sum):

$$U = 1/(R_1 + R_2 + R_3 +...+ R_n)$$

Example #1

The plan section shown on page 50 is for a wall being considered for a heat loss condition (winter). What is the U value of the wall?

Solution:

We tabulate the resistances in column format. We will calculate the U value both at a stud and between the studs. We will then calculate the weighted average, which will represent the U value of the wall as a whole. The air film adjacent to the wall on both sides is always

TABLE 3.3 — THERMAL RESISTANCES

Description	Density lb/ft³	Conductivity[b] (k) Btu•in./ h•ft²•F	Conductance (C) Btu/h• ft²•F	Resistance [c] (R)		Specific Heat Btu/lb • deg F
				Per inch thickness (1/k) h•ft²• F/Btu	For thickness listed (1/C) h•ft²• F/Btu	
INSULATING MATERIALS						
Blanket and Batt[d]						
Mineral Fiber, fibrous form processed from rock, slag, or glass						
approx.[e] 3–4 in.	0.3–2.0	—	0.091	—	11[d]	
approx.[e] 3.5 in.	0.3–2.0	—	0.077	—	13[d]	
approx.[e] 5.5–6.5 in.	0.3–2.0	—	0.053	—	19[d]	
approx.[e] 6–7.5 in.	0.3–2.0	—	0.045	—	22[d]	
approx.[e] 9–10 in.	0.3–2.0	—	0.033	—	30[d]	
approx.[e] 12–13 in.	0.3–2.0	—	0.026	—	38[d]	
MASONRY MATERIALS						
Concretes						
Cement mortar	116	5.0	—	0.20	—	
Gypsum-fiber concrete 87.5% gypsum, 12.5% wood chips	51	1.66	—	0.60	—	0.21
Lightweight aggregates including ex-	120	5.2	—	0.19	—	
panded shale, clay or slate; expanded	100	3.6	—	0.28	—	
slags; cinders; pumice; vermiculite;	80	2.5	—	0.40	—	
also cellular concretes	60	..7	—	0.59	—	
	40	1.15	—	0.86	—	
	30	0.90	—	1.11	—	
	20	0.70	—	1.43	—	
Perlite, expanded	40	0.93	—	1.08	—	
	30	0.71	—	1.41	—	
	20	0.50	—	2.00	—	0.32
Sand and gravel or stone aggregate (oven dried)	140	9.0	—	0.11	—	0.22
Sand and gravel or stone aggregate (not dried)	140	12.0	—	0.08	—	
Stucco	116	5.0	—	0.20	—	
WOODS (12% Moisture Content)[o,p]						
Hardwoods						0.39
Oak	41.2–46.8	1.12-1.25	—	0.89-0.80	—	
Birch	42.6–45.4	1.16-1.22	—	0.87-0.82	—	
Maple	39.8–44.0	1.09-1.19	—	0.94-0.88	—	
Ash	38.4–41.9	1.06-1.14	—	0.94-0.88	—	
Softwoods						0.39
Southern Pine	35.6–41.2	1.00-1.12	—	1.00-0.89	—	
Douglas Fir-Larch	33.5–36.3	0.95-1.01	—	1.06-0.99	—	
Southern Cypress	31.4–32.1	0.90-0.92	—	1.11-1.09	—	
Hem-Fir, Spruce-Pine-Fir	24.5–31.4	0.74-0.90	—	1.35-1.11	—	
West Coast Woods, Cedars	21.7–31.4	0.68-0.90	—	1.48-1.11	—	
California Redwood	24.5–28.0	0.74-0.82	—	1.35-1.22	—	
PLASTERING MATERIALS						
Cement plaster, sand aggregate	116	5.0	—	0.20	—	0.20
Sand aggregate ... 0.375 in.	—	—	13.3	—	0.08	0.20
Sand aggregate ... 0.75 in.	—	—	6.66	—	0.15	0.20
Gypsum plaster:						
Lightweight aggregate ... 0.5 in.	45	—	3.12	—	0.32	
Lightweight aggregate ... 0.625 in.	45	—	2.67	—	0.39	
Lightweight agg. on metal lath ... 0.75 in.	—	—	2.13	—	0.47	
Perlite aggregate	45	1.5	—	0.67	—	0.32

considered as a vertical layer because of its resistance to heat flow. The orientation of air films and air space is vertical and the heat flow is horizontal. The winter case always assumes a 15 mph wind outside and a low average temperature in the wall. Since the air space is not the

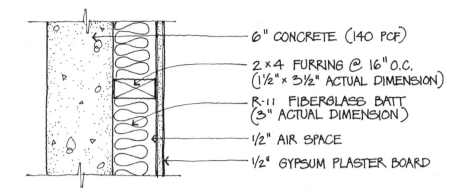

6" CONCRETE (140 PCF)

2×4 FURRING @ 16" O.C.
(1½" × 3½" ACTUAL DIMENSION)

R-11 FIBERGLASS BATT
(3" ACTUAL DIMENSION)

½" AIR SPACE

½" GYPSUM PLASTER BOARD

CALCULATION OF U VALUE

dominant resistance in the wall (the insulation is) the temperature drop across the air space will be small. The emissivity on each side of the air space is high because we assume there is only building paper on the fiberglass, and nothing special on the back of the gypsum board. If there were foil backing on both, the emissivity would be low.

Note: All of the following values come from Tables 3.2 and 3.3.

	R@ gap	R@ stud
Outside air film (assume wind = 15 mph)	.17	.17
Concrete (k = 9, x = 6" x/k = 6/9 = .67)	.67	.67
Stud (x/k = 3.5/1.0 = 3.5)	–	3.5
Fiberglass (glass batt)	11.00	–
Air space (50°F mean temp, 30°ΔT, ε = .82)	.90	–
1/2" gypsum board	.32	.32
Inside air film (still air, ε = .90)	.68	.68
$R_{tot} = \Sigma R =$	13.74	5.34
$U = 1/\Sigma R =$	.073	.187

Finding a weighted average for the U value is based on 1 1/2" of .187 (stud) and 14 1/2" of .073 (insulated gap) for every 16" of wall.

$$\frac{1.5(.187) + 14.5(.073)}{16} = 0.08366$$
$$= 0.08\ Btuh/ft^2°F$$

Although this problem is more complicated than any single problem you are likely to encounter on the exam, it serves to illustrate a number of factors we have discussed. Such a problem might appear on the exam broken up into several problems.

Latent Heat

When you get warm enough, you begin to perspire. The sweat evaporates, cooling the body. That is also a form of heat transfer caused by the change of state of the perspiration from the liquid state (water) to the vapor state (humidity). This is called a *phase change*, and either stores ("uses up") or releases energy. In the case of evaporation, it "uses up" energy, which was excess to the body *(latent heat of evaporation).* Ice melting in a glass of ice water "uses up" heat, keeping the drink cool *(latent heat of fusion).* If we wanted to recreate the ice cube, all of that energy would have to be extracted from the water again (by refrigeration) so in that sense, the energy is *stored* in the water. It is now becoming typical in solar designs to store energy using phase change materials such as eutectic salts (which dissolve or crystallize in water) or special paraffins that melt or solidify at comparatively low temperatures.

This stores and releases solar energy (heat) without tremendous temperature variations, which is vastly preferable in buildings.

Removing the moisture from the air in hot humid buildings always requires a great deal of energy, because of the heat that must be extracted to get the moisture to condense out (change state). You may remember that it takes 1 Btu to change 1 pound of water 1°F. How many Btu's does it take to get 1 pound of water from freezing to boiling?

$$212°F - 32°F = 180°F$$

$$180°F \times 1 \text{ Btu/°Ib} = \textit{180 Btu}$$

Compare this with the amount it takes to boil (or evaporate) the same pound of water at 212°F into one pound of steam, still at 212°F. The latent heat of evaporation is 1,000 Btu's/pound for water.

$$1 \text{ lb. } H_2O_{liq} \text{ @ } 212°F + \textit{1,000 Btu}$$

$$=> 1 \text{ lb. } H_2O_{vap} \text{ @ } 212°F$$

It takes 1,000 Btu's just to evaporate a pound of water compared to 180 Btu's for the entire temperature change from freezing to boiling!

HEATING LOAD CALCULATIONS

The heating load of a building consists of the sum of all of the losses through the building skin. Given an understanding of the basic heat transfer processes, how do we apply these concepts to an actual building?

Conduction (q_c or HL_c)

The same basic formula is used for conduction whether it occurs through the roof, walls, windows or doors. The heat flow is the product of the conductance of the assembly (expressed by the U value), the temperature differential between inside and outside (ΔT), and the exposed surface area of wall, window, or roof (A). The instantaneous version of this formula, which measures the flow at a given instant in time, is:

$$q_c = U \ (A)\Delta T$$

$$= U \ (A)(T_{inside} - T_{outside})$$

The conduction is expressed in Btu's per hour, which is a measure of energy flow rate per unit of time at any given instant. If that flow rate were to continue for an hour, the given number of Btu's would be transferred. This is often abbreviated as Btu/h or Btuh.

A similar formula is used for an energy flow rate over a long period of time:

$$q_c = U \ (A)24(DD)$$

where DD refers to a number of *degree days* and q_c refers to total Btu's. A degree day is a measure of how cold it has been at a given place over a given period of time. One degree day is defined as a day whose mean temperature is one degree below the reference temperature of 65°F. Two degree days may be a single day with a mean temperature of 63° (two degrees below 65°) or two days at 64°. The amount of energy necessary to heat the building is assumed to be the same in either case. All days with a mean temperature above 65°F are disregarded. The idea is that we can record the temperature and calculate the degree days for a particular locality over a period such as a month and calculate the total heat loss that would result. We can even keep records for an entire winter. A mild winter would be a 3,000 degree day winter, while a severe winter would be a 7,000 degree day winter.

The instantaneous version of q_c is used to determine the case at a particular moment, usually called the *design day*. This is a day colder than 98 percent of the days experienced in that climate. If the heating equipment and plant are sized to keep the building warm on that day, they would be sufficient for the other 98 percent as well.

The degree day version of the formula for q_c may be used to compare two alternatives over a longer period of time. Thus, it is possible to determine a *payback* period on an investment, for example, the number of years of the reduced energy costs it would take to pay for the increase in insulation.

Example #2

If the design day for Hepzibah, NY, is 10°F, and if we expect to maintain an interior temperature of 65°F, what will be the conducted heat loss through 200 ft² of the wall section of Example #1?

Solution:

The calculated U value of Example #1 was .08 Btuh/ft²°F.

$$q_c = U(A)\Delta T$$
$$= .08 \text{ Btuh/ft}^2°F \times (200\text{ft}^2)$$
$$\times (65°F - 10°F)$$
$$= \textit{880 Btuh}$$

Furthermore, if Hepzibah experienced a 6,600 degree day winter, how much heat would be lost through the wall that year?

$$q_c = U(A)24 \text{ DD}$$
$$= .08 \text{ Btuh/ft}^2°F \times (200 \text{ ft}^2)$$
$$\times 24 \text{ hr} \times 6,600 \text{ DD}$$
$$= \textit{2,534,000 Btu} \text{ or about}$$
$$\textit{2.5 million Btu's}$$

Conduction below Grade

Although q_c theoretically applies to building elements below the ground surface, it becomes very difficult to determine what value to use for the outside temperature, since it varies with depth and moisture. It is therefore difficult to determine a reasonable value for ΔT. In addition, the loss through basement walls and floor slabs is comparatively low. The values for *below grade walls* are therefore taken from a table based on ground water temperature, which is usually assumed to be the same as the average annual air temperature. The *slab on grade* values are taken from a table which considers whether or not the slab edge is insulated. The total loss (q_s) is simply the area times the factor taken from the table.

Infiltration

All buildings leak air. There is a steady flow of air in and out, through cracks such as those between window sash and frame, and even through walls where there are sockets and switches or (heaven forbid) where there is sloppy construction.

The outside air that leaks in replaces internal air and must therefore be heated or cooled from the outside temperature to the desired inside temperature. The heat required is called the infiltration load (q_i). It is calculated in two steps:

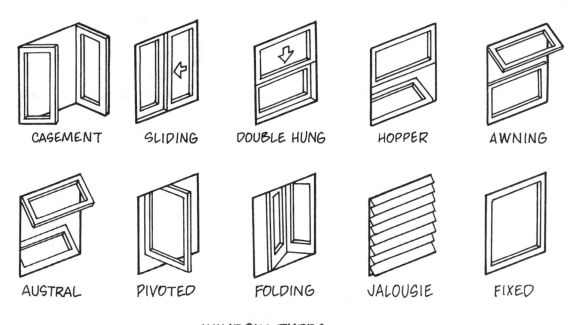

WINDOW TYPES VIEWED FROM OUTSIDE

1. Determine the amount of air infiltration.
2. Determine the amount of heating (or cooling) required to bring the air to the proper temperature.

The amount of air infiltration may by determined using the *air change method* or the *crack method*. The air change method requires that you know the number of air changes per hour in the building. For existing buildings this may be determined by measurement, but for new buildings this must be estimated. The air change method is also useful for very tight buildings, such as offices, where there might not be much infiltration, but which require some minimum number of air changes per hour anyway, for hygiene or code reasons. The amount of air (Q_{cfh}, expressed in cubic feet per *hour*) is determined by multiplying the building volume in cubic feet (V) by the number of air changes (N):

$$Q_{cfh} = N \times V$$

The crack method is based on the number of linear feet of crack or joint in all of the windows in the space under consideration, which might be one room or the entire building.

For example, a 3 foot × 6 foot window has 3 ft + 6 ft + 3 ft + 6 ft = 18 feet of crack. If it were a double hung window, which has a joint in the middle as well, that would add 3 feet for a total of 21 feet. The amount of infiltration per linear foot is determined from a table which considers wind speed and window type. This value may then be multiplied by the number of linear feet:

$$Q_{cfh} = LF \times CFH/lin.ft.$$

Finally, the amount of heating or cooling required may be calculated using the equation:

$$q_i = .018 \, (Q_{cfh})\Delta T$$
$$= .018 \, (Q_{cfh})(T_{inside} - T_{outside})$$

Total Heating Load

We now have all of the pieces and can combine them to calculate the total heating load:

$$q_{total} = q_c + q_s + q_i$$

Note that q_c may be composed of several different "sub" q_c's, one for each surface, i.e., q_c for the walls, q_c for the windows and q_c for the roof. You will often have to break up q_c for the walls into one q_c for each type of wall section, since each wall section may have a different U value.

TEMPERATURE GRADIENTS

While the above formulas tell us what is happening to the inside of a building, they do not tell us what is happening to the inside of a wall. For example, why do pipes within walls freeze and burst when the room temperature is 65°F? The temperature of each point within a wall depends on the resistance of the layers up to that point compared with the total resistance of all the materials in the wall. For example, though the room temperature may be 65°F, the temperature inside part of the wall may be below 32°F. The equation for the temperature drop across a layer is:

$$\Delta T_{layer} = (R_{layer}/R_{total}) \times \Delta T_{total}$$

Therefore we can calculate the temperature at the boundary between two layers by calculating the sum of the temperature gradients to that point.

Example #3

Determine the temperatures inside the walls in Hepzibah, NY, on the same design day as in Example #2. Investigate only the section through the air space, since there are water pipes there, which should not be allowed to freeze.

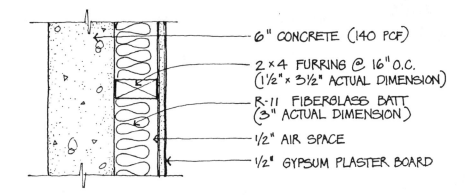

6" CONCRETE (140 PCF)

2 × 4 FURRING @ 16" O.C.
(1½" × 3½" ACTUAL DIMENSION)

R-11 FIBERGLASS BATT
(3" ACTUAL DIMENSION)

½" AIR SPACE

½" GYPSUM PLASTER BOARD

Solution:

	R	ΔT	T	**Boundary of** **Outside air**
			10.00°F	
Outside air film	.17	$(.17/13.74) \times 55° = 0.68$		Air film
			10.68°F----------------	
Concrete	.67	$(.67/13.74) \times 55° = 2.68$		Concrete
			13.36°F----------------	
Fiberglass	11.00	$(11.0/13.74) \times 55° = 44.03$		Fiberglass
			57.39°F----------------	
Air space	.90	$(.90/13.74) \times 55° = 3.6$		Air space
			60.99°F----------------	
Gypsum board	.32	$(.32/13.74) \times 55° = 1.29$		Gyp.brd.
			62.28°F----------------	
Inside air film	.68	$(.68/13.74) \times 55° = 2.72$		Air film
			65.00°F	
				Inside air

COOLING LOAD CALCULATIONS

There are a number of internal sources of heat in a building, which should be considered when sizing cooling equipment, and when calculating the conducted transfer through the walls. This is complicated since several heat transfer processes are involved simultaneously. (See CLTD.)

People (q_p)

The occupants of a building themselves comprise one of the sources of heat gain. This can range from a minor factor (two or three people in a house) to a dominant constraint (3,000 people in an auditorium). Not only is the number important, but so is the activity. A human being generates about 450 Btuh at rest, but as much as 2,500 Btuh when engaged in heavy work or athletic activity. The cooling load may be calculated by looking up the activity and multiplying by the appropriate number of people:

$$q_p = \text{No. of people} \times \text{Btuh/person}$$

TABLE 3.6 – BTUH FOR VARIOUS ACTIVITIES

Degree of Activity	Typical Application	Total Heat Adults, Male Btu/h	Total Heat Adjusted[b] Btu/h	Sensible Heat Btu/h	Latent Heat Btu/h
Seated at rest	Theater, movie	400	350	210	140
Seated, very light work writing	Offices, hotels, apts	480	420	230	190
Seated, eating	Restaurant[c]	520	580 [c]	255	325
Seated, light work, typing	Offices, hotels, apts	640	510	255	255
Standing, light work or walking slowly	Retail Store, bank	800	640	315	325
Light bench work	Factory	880	780	345	435
Walking, 3 mph, light machine work	Factory	1040	1040	345	695
Bowling[d]	Bowling alley	1200	960	345	615
Moderate dancing	Dance hall	1360	1280	405	875
Heavy work, heavy machine work, lifting	Factory	1600	1600	565	1035
Heavy work, athletics	Gymnasium	2000	1800	635	1165

Lighting (q_l)

Lighting equipment generates light, of course. But it also generates heat. In fact, incandescent lighting generates much more heat than light, and even light itself eventually gets absorbed and turns into heat. Thus, even efficient fixtures generate heat, directly in proportion to their wattage. This heat must be removed by the cooling equipment like any other load. The equation for the load is:

$$q_l = 3.4 \ W$$

where W = the wattage of the equipment.

Equipment (q_m)

Mechanical and electrical equipment produce heat as well, and the cooling load can be determined in one of several ways, depending on the information available. Sometimes the Btuh is part of the equipment specification, such as with a stove or heater. Sometimes only the wattage is available (typewriter, hair dryer, or

TV). In such cases, use the formula for q_l on the previous page. In some cases, the only information available is the horsepower of the equipment (i.e., a diesel generator). Most equipment denominated in horsepower does not run at full horsepower most of the time, and thus a discounting factor is applied. The commonly accepted equation is:

$$q_m = 1,500 \times Bhp$$

where Bhp is brake horsepower (the most common measurement). One Bhp is actually 2,545 Btuh.

Cooling Load Temperature Differential (CLTD or ETD)

Calculating heat gain through walls is complicated by the fact that peak gains usually occur in sunny weather, and often occur when there is a large diurnal (day to night to day) temperature swing. Under such circumstances, the thermal mass and storage capacity of the wall, as well as its color and orientation, are important factors (see the discussion of C_p and ε on pages 44 and 45). The formulas for q_c on page 51 do not take any of these factors into account, yet it is difficult to separate them into individual equations. As a result, they are often lumped together to get an "equivalent" temperature differential to use in the q_c equation. Thus the new equation is:

$$q_{CLTD} = U\ (A)\ CLTD$$

where CLTD is Cooling Load Temperature Differential, or:

$$q_{ETD} = U\ (A)\ ETD$$

where ETD is Equivalent Temperature Differential.

The complications arise in determining CLTD. There are three steps:

1. Determine the wall classification or group of the wall or roof section.

2. Determine the base CLTD from the time of day, wall type, wall orientation, and wall or roof color.

3. Adjust the CLTD based on the temperature history of the last 24 hours.

All of the steps are straightforward except the last. The tables are based on an average *outside* temperature of 85°F (i.e., a temperature swing of 75 to 95°F). If the average temperature was 8 degrees higher (say the swing was 86 to 100°F for an average of 93°F), then you would have to add 8 degrees to the CLTD. Similarly, the assumed *interior* temperature is 78°F, and if you had set the thermostat at a cooler temperature, you would have to increase the CLTD by the appropriate amount.

Example #4

Let's take our Hepzibah, NY, wall and find the peak heat gain on an August day in which the average outdoor temperature during the preceding 24 hours was 83°F. Assume that the wall faces south, and that it is painted a dark color. What is the peak heat gain and when does it occur?

Solution:

We don't know when we will experience the greatest CLTD, because we don't yet know how long the thermal mass will delay the peak load.

1. Find the wall group from Table 3.7. The 8" concrete with 3" of insulation is not shown in the table, but there is a listing for 8" concrete with 2" of insulation in group A. Since we have more insulation than the wall listed, we will use the U value of 0.08 calculated in Example #1, rather than the tabulated value. The mass of the listed wall, however, is very close to that of the wall in the problem.

2. From Table 3.8, we see that the peak CLTD is 20°F and occurs at 11:00 p.m.

3. The average outdoor temperature over the preceding 24 hours was given as 83°F, which is 2° below the reference temperature of 85°F for the CLTD table. We must therefore lower the CLTD by 2°, resulting in a CLTD of 18°F.

4. The peak gain occurs at 11:00 p.m. and is:

$$q_{CLTD} = U\ (A)\ CLTD$$
$$= .08\ Btuh/ft^2°F \times 200\ ft^2 \times 18°F$$
$$= \textit{288 Btuh}$$

TABLE 3.7 — CLTD WALL GROUPS

Group No.	Description of Construction	Weight (lb/ft²)	U-Value (Btu/h·ft²·F)	Code Numbers of Layers (see Table 8)
4-in. Face Brick + (*Brick*)				
C	Air Space + 4-in. Face Brick	83	0.358	A0, A2, B1, A2, E0
D	4-in. Common Brick	90	0.415	A0, A2, C4, E1, E0
C	1-in. Insulation or Air Space + 4-in. Common Brick	90	0.174-0.301	A0, A2, C4, B1/B2, E1, E0
B	2-in. Insulation + 4-in. Common Brick	88	0.111	A0, A2, B3, C4, E1, E0
B	8-in. Common Brick	130	0.302	A0, A2, C9, E1, E0
A	Insulation or Air Space + 8-in. Common brick	130	0.154-0.243	A0, A2, C9, B1/B2, E1, E0
4-in. Face Brick + (*H.W. Concrete*)				
C	Air Space + 2-in. Concrete	94	0.350	A0, A2, B1, C5, E1, E0
B	2-in. Insulation + 4-in. Concrete	97	0.116	A0, A2, B3, C5, E1, E0
A	Air Space or Insulation + 8-in. or more Concrete	143-190	0.110-0.112	A0, A2, B1, C10/11, E1, E0
4-in. Face Brick + (*L.W. or H.W. Concrete Block*)				
E	4-in. Block	62	0.319	A0, A2, C2, E1, E0
D	Air Space or Insulation + 4-in. Block	62	0.153-0.246	A0, A2, C2, B1/B2, E1, E0
D	8-in. Block	70	0.274	A0, A2, C7, A6, E0
C	Air Space or 1-in. Insulation + 6-in. or 8-in. Block	73-89	0.221-0.275	A0, A2, B1, C7/C8, E1, E0
B	2-in. Insulation + 8-in. Block	89	0.096-0.107	A0, A2, B3, C7/C8, E1, E0
4-in. Face Brick + (*Clay Tile*)				
D	4-in. Tile	71	0.381	A0, A2, C1, E1, E0
D	Air Space + 4-in. Tile	71	0.281	A0, A2, C1, B1, E1, E0
C	Insulation + 4-in. Tile	71	0.169	A0, A2, C1, B2, E1, E0
C	8-in. Tile	96	0.275	A0, A2, C6, E1, E0
B	Air Space or 1-in. Insulation + 8-in. Tile	96	0.142-0.221	A0, A2, C6, B1/B2, E1, E0
A	2-in. Insulation + 8-in. Tile	97	0.097	A0, A2, B3, C6, E1, E0
H.W. Concrete Wall + (*Finish*)				
E	4-in. Concrete	63	0.585	A0, A1, C5, E1, E0
D	4-in. Concrete + 1-in. or 2-in. Insulation	63	0.119-0.200	A0, A1, C5, B2/B3, E1, E0
C	2-in. Insulation + 4-in. Concrete	63	0.119	A0, A1, B6, C5, E1, E0
C	8-in. Concrete	109	0.490	A0, A1, C10, E1, E0
B	8-in. Concrete + 1-in. or 2-in. Insulation	110	0.115-0.187	A0, A1, C10, B5/B6, E1, E`
A	2-in. Insulation + 8-in. Concrete	110	0.115	A0, A1, B3, C10, E1, E0
B	12-in. Concrete	156	0.421	A0, A1, C11, E1, E0
A	12-in. Concrete + Insulation	156	0.113	A0, C11, B6, A6, E0
L.W. and H.W. Concrete Block + (*Finish*)				
F	4-in. Block + Air Space/Insulation	29	0.161-0.263	A0, A1, C2, B1/B2, E1, E0
E	2-in. Insulation + 4-in. Block	29-37	0.105-0.114	A0, A1, B3, C2/C3, E1, E0
E	8-in. Block	47-51	0.294-0.402	A0, A1, C7/C8, E1, E0
D	8-in. Block + Air Space/Insulation	41-57	0.149-0.173	A0, A1, C7/C8, B1/B2, E1, E0
Metal Curtain Wall				
G	With/without air Space + 1-in./ 2-in. 3-in. Insulation	5-6	0.091-0.2 30	A0, A3, B5/B6/B12, A3, E0
Frame Wall				
G	1-in. to 3-in. Insulation	16	0.081-0.1 78	A0, A1, B1, B2/B3/B4, E1, E0

RADIATION THROUGH WINDOWS (q_r OR SHGF)

There are several methods for calculating the radiant gain through glass, also known as *insolation* (not to be confused with insulation).

CLTD does not really apply, since the glass transmits radiation and conducts heat separately, because it has neither time lag nor thermal mass.

The formula for q_c can be used for the conducted gain. Radiation can be calculated by finding the

TABLE 3.8 – CLTD VALUES

North Latitude Wall Facing	0100	0200	0300	0400	0500	0600	0700	0800	0900	1000	1100	1200	1300	1400	1500	1600	1700	1800	1900	2000	2100	2200	2300	2400	Hr of Maximum CLTD	Minimum CLTD	Maximum CLTD	Difference CLTD
Group A Walls																												
N	14	14	14	13	13	13	12	12	11	11	10	10	10	10	10	10	11	11	12	12	13	13	14	14	2	10	14	4
NE	19	19	19	18	17	17	16	15	15	15	15	15	16	16	17	18	18	18	19	19	20	20	20	20	22	15	20	5
E	24	24	23	23	22	21	20	19	19	18	19	19	20	21	22	23	24	24	25	25	25	25	25	25	22	18	25	7
SE	24	23	23	22	21	20	20	19	18	18	18	18	18	19	20	21	22	23	23	24	24	24	24	24	22	18	24	6
S	20	20	19	19	18	18	17	16	16	15	14	14	14	14	14	15	16	17	18	19	19	20	20	20	23	14	20	6
SW	25	25	25	24	24	23	22	21	20	19	19	18	17	17	17	17	18	19	20	22	23	24	25	25	24	17	25	8
W	27	27	26	26	25	24	24	23	22	21	20	19	19	18	18	18	18	19	20	22	23	25	26	26	1	18	27	9
NW	21	21	21	20	20	19	19	18	17	16	16	15	15	14	14	14	15	15	16	17	18	19	20	21	1	14	21	7
Group B Walls																												
N	15	14	14	13	12	11	11	10	9	9	9	8	9	9	9	10	11	12	13	14	14	15	15	15	24	8	15	7
NE	19	18	17	16	15	14	13	12	12	13	14	15	16	17	18	19	19	20	20	21	21	21	20	20	21	12	21	9
E	23	22	21	20	18	17	16	15	15	15	17	19	21	22	24	25	26	26	27	27	26	26	25	24	20	15	27	12
SE	23	22	21	20	18	17	16	15	14	14	15	16	18	20	21	23	24	26	26	26	25	24	23	21	19	14	26	12
S	21	20	19	18	17	15	14	13	12	11	11	11	11	12	14	15	17	19	20	21	22	22	21	23	23	11	22	11
SW	27	26	25	24	22	21	19	18	16	15	14	14	13	13	14	15	17	20	22	25	27	28	28	28	24	13	28	15
W	29	28	27	26	24	23	21	19	18	17	16	15	14	14	15	17	19	22	25	27	29	30	30	29	22	14	30	16
NW	23	22	21	20	19	18	17	15	14	13	12	12	12	11	12	12	13	15	17	19	21	22	23	24	24	11	23	9
Group C Walls																												
N	15	14	13	12	11	10	9	8	8	7	7	8	8	9	10	12	13	14	15	16	17	17	17	16	22	7	17	10
NE	19	17	16	14	13	11	10	10	11	13	15	17	19	20	21	22	22	23	23	23	23	21	20	20	20	10	23	13
E	22	21	19	17	15	14	12	12	14	16	19	22	25	27	29	29	30	30	30	29	28	27	26	24	18	12	30	18
SE	22	21	19	17	15	14	12	12	12	13	16	19	22	24	26	28	29	29	29	29	28	27	26	24	19	12	29	17
S	21	19	18	16	15	13	12	10	9	9	9	10	11	14	17	20	22	24	25	26	25	24	22	20	19	9	26	17
SW	29	27	25	22	20	18	16	15	13	12	11	11	11	13	15	18	22	26	29	32	33	33	32	31	21	11	33	22
W	31	29	27	25	22	20	18	16	14	13	12	12	12	13	14	16	20	24	29	32	35	35	35	33	22	12	35	23
NW	25	23	21	20	18	16	14	13	11	10	10	10	10	11	12	13	15	18	22	25	27	27	27	26	22	10	27	17
Group D Walls																												
N	15	13	12	10	9	7	6	6	6	6	6	7	8	10	12	13	15	17	18	19	19	19	18	16	21	6	19	13
NE	17	15	13	11	10	8	7	8	10	14	17	20	22	23	23	24	24	25	25	24	23	22	20	18	19	7	25	18
E	19	17	15	13	11	9	8	9	12	17	22	27	30	32	33	33	32	32	31	30	28	26	24	22	16	8	33	25
SE	20	17	15	13	11	10	8	8	10	13	17	22	26	29	31	32	32	32	31	30	28	26	24	22	17	8	32	24
S	19	17	15	13	11	9	8	7	6	6	7	9	12	16	20	24	27	29	29	29	27	26	24	22	19	6	29	23
SW	28	25	22	19	16	14	12	10	9	8	8	8	10	12	15	21	27	33	38	38	37	34	31	21	19	8	38	30
W	31	27	24	21	18	15	13	11	10	9	9	9	10	12	14	18	24	30	36	40	41	40	38	34	21	9	41	32
NW	25	22	19	17	14	12	10	9	8	7	7	8	9	10	12	14	18	22	27	31	32	32	30	27	22	7	32	25
Group E Walls																												
N	12	10	8	7	5	4	3	4	5	6	7	9	11	13	15	17	19	20	21	23	20	18	16	14	20	3	22	19
NE	13	11	9	7	6	4	5	9	15	20	24	25	25	26	26	26	26	26	25	24	22	19	17	15	16	4	26	22
E	14	12	10	8	6	5	6	11	18	26	33	36	38	37	36	34	33	32	30	28	25	22	20	17	13	5	38	33
SE	15	12	10	8	7	5	5	8	12	19	25	31	35	37	37	36	34	33	31	28	26	23	20	17	15	5	37	32
S	15	12	10	8	7	5	4	3	4	5	9	13	19	24	29	32	34	33	31	29	26	23	20	17	16	3	34	31
SW	22	18	15	12	10	8	6	5	5	6	7	9	12	18	24	32	38	43	45	44	40	35	30	26	19	5	45	40
W	25	21	17	14	11	9	7	6	6	6	7	9	11	14	20	27	36	43	49	49	45	40	34	29	19	6	49	43
NW	20	17	14	11	9	7	6	5	5	5	6	8	10	13	16	20	26	32	37	38	36	32	28	24	20	5	38	33
Group F Walls																												
N	8	6	5	3	2	1	2	4	6	7	9	11	14	17	19	21	22	23	24	23	20	16	13	11	19	1	23	23
NE	9	7	5	3	2	1	5	14	23	28	30	29	28	27	27	26	24	22	19	16	13	11	11	10	11	1	30	29
E	10	7	6	4	3	2	6	17	28	38	44	45	43	39	36	34	32	30	27	24	21	17	15	12	12	2	45	43
SE	10	7	6	4	3	2	4	10	19	28	36	41	43	42	39	36	34	31	28	25	21	18	15	12	13	2	43	41
S	10	8	6	4	3	2	1	1	3	7	13	20	27	34	38	39	38	35	31	26	22	18	15	12	16	1	39	38
SW	15	11	9	6	5	3	2	2	4	5	8	11	17	26	35	44	50	53	52	45	37	28	23	18	18	2	53	48
W	17	13	10	7	5	4	3	3	4	6	8	11	14	20	28	39	49	57	60	54	43	34	27	21	19	3	60	57
NW	14	10	8	6	4	3	2	2	3	5	8	10	13	15	21	27	35	42	46	43	35	28	22	18	19	2	46	44
Group G Walls																												
N	3	2	1	0	−1	2	7	8	9	12	15	18	21	23	24	24	25	26	22	15	11	9	7	5	18	−1	26	27
NE	3	2	1	0	−1	9	27	36	39	35	30	26	26	27	27	26	25	22	18	14	11	9	7	5	9	−1	39	40
E	4	2	1	0	−1	11	31	47	54	55	50	40	33	31	30	29	27	24	19	15	12	10	8	6	10	−1	55	56
SE	4	2	1	0	−1	5	18	32	42	49	51	48	42	36	32	30	27	24	19	15	12	10	8	6	11	−1	51	52
S	4	2	1	0	−1	0	1	5	12	22	31	39	45	46	43	37	31	25	20	15	12	10	8	6	14	−1	46	47
SW	5	4	3	1	0	2	5	8	12	16	26	38	50	59	63	61	52	37	24	17	13	10	8	6	15	0	63	63
W	6	5	3	2	1	1	2	5	8	11	15	19	27	41	56	67	72	67	48	29	20	15	11	8	17	1	72	71
NW	5	3	2	1	0	0	2	5	8	11	15	18	21	27	37	47	55	55	41	25	17	13	10	7	18	0	55	55

intensity of the direct sun on the surface of the glass, and then multiplying by the glass area and the percentage transmitted. The percentage transmitted compared to that which is transmitted by clear glass is called the *shading coefficient (SC)* and is somewhat similar to the transmissivity. However, half of the radiation absorbed is assumed to reradiate into the space, as well. Therefore the tabulated shading coefficient values are different from those for

TABLE 3.9 — SHADING COEFFICIENTS

A. Single Glass

Type of Glass	Nominal Thickness[b]	Solar Trans.[b]	Shading Coefficient h_0=4.0	Shading Coefficient h_0=3.0
Clear	1/8 in.	0.86	1.00	1.00
	1/4 in.	0.78	0.94	0.95
	3/8 in.	0.72	0.90	0.92
	1/2 in.	0.67	0.87	0.88
Heat Absorbing	1/8 in.	0.64	0.83	0.85
	1/4 in.	0.46	0.69	0.73
	3/8 in.	0.33	0.60	0.64
	1/2 in.	0.24	0.53	0.58

B. Insulating Glass

Clear Out, Clear In	1/8 in.[c]	0.71[e]	0.88	0.88
Clear Out, Clear In	1/4 in.	0.61	0.81	0.82
Heat Absorbing[d] Out, Clear In	1/4 in.	0.36	0.55	0.58

[a] Refers to factory-fabricated units with 3/16, 1/4 or 1/2-in. air space or to prime windows plus storm sash.
[b] Refer to manufacturer's literature for values.
[c] Thickness of each pane of glass, not thickness of assembled unit.
[d] Refers to gray, bronze and green tinted heat-absorbing float glass.
[e] Combined transmittance for assembled unit.

transmissivity (see Table 3.9). Generally speaking, clear glass lets in a lot of light and heat, tinted glass lets in less light and some heat, and reflective glass lets in less light and less heat. Unfortunately, it is not always desirable to let in less light, because more artificial lighting may be required, which in turn generates more heat. It is best to stop the *direct* sunlight *outside* the building, using fins, overhangs, or awnings. The diffuse daylight which remains provides natural lighting and contributes much less heat. If direct sunlight is blocked inside the glass, with a Venetian blind or a drape, all of the heat that gets absorbed has already been admitted to the space. If it is necessary to use internal shielding, the more reflective or light colored the better.

The formula for radiant gain (sometimes called *solar heat gain factor* or just *solar factor*) is:

$$q_r (= SHGF = SF) = S_g (SC) A$$

where

S_g = the intensity (in Btuh/ft^2) on a surface area in a given orientation

SC = the shading coefficient

A = the area exposed to direct sunlight

We will discuss solar gain further in the next lesson.

TABLE 3.10 — SOLAR INTENSITY (S_G) FOR 40° NORTH LATITUDE

Date	Solar Time	Direct Normal Btuh•ft²	N	NNE	NE	ENE	E	ESE	SE	SSE	S	SSW	SW	WSW	W	WNW	NW	NNW	HOR	Solar Time
Mar 21	0700	171	9	29	93	140	163	161	135	86	22	8	8	8	8	8	8	8	26	1700
	0800	250	16	18	91	169	218	232	211	157	74	17	16	16	16	16	16	16	85	1600
	0900	282	21	22	47	136	203	238	236	198	128	40	22	21	21	21	21	21	143	1500
	1000	297	25	25	27	72	153	207	229	216	171	95	29	25	25	25	25	25	186	1400
	1100	305	28	28	28	30	78	151	198	213	197	150	77	30	28	28	28	28	213	1300
	1200	307	29	29	29	29	31	75	145	191	206	191	145	75	31	29	29	29	223	1200
	HALF DAY TOTALS		114	139	302	563	832	1035	1087	968	694	403	220	132	114	113	113	113	764	
Jun 21	0500	22	10	17	21	22	20	14	6	2	1	1	1	1	1	1	1	2	3	1900
	0600	155	48	104	143	159	151	121	70	17	13	13	13	13	13	13	14	40	1800
	0700	216	37	113	172	205	207	178	122	46	22	21	21	21	21	21	21	97	1700
	0800	246	30	85	156	201	216	199	152	80	29	27	27	27	27	27	27	153	1600
	0900	263	33	51	114	166	192	190	161	105	45	33	32	32	32	32	32	201	1500
	1000	272	35	38	63	109	145	158	148	116	69	39	36	35	35	35	35	238	1400
	1100	277	38	39	40	52	81	105	116	110	88	60	41	39	38	38	38	260	1300
	1200	279	38	38	38	40	41	52	72	89	95	89	72	52	41	40	38	38	267	1200
	HALF DAY TOTALS		253	470	734	941	1038	999	818	523	315	236	204	191	188	187	186	188	1126	
Dec 21	0800	89	3	3	8	41	67	82	84	73	50	17	3	3	3	3	3	3	6	1600
	0900	217	10	10	11	60	135	185	205	194	151	83	13	10	10	10	10	10	39	1500
	1000	261	14	14	14	25	113	188	232	239	210	146	55	15	14	14	14	14	77	1400
	1100	280	17	17	17	17	56	151	217	249	242	198	120	28	17	17	17	17	104	1300
	1200	285	18	18	18	18	19	89	178	233	253	233	178	89	19	18	18	18	113	1200
	HALF DAY TOTALS		52	52	56	146	374	649	822	867	775	557	276	94	53	52	52	52	282	

Solar Heat Gain Factors, Btuh • ft²

SUMMARY

In this lesson we have discussed the different modes of heat transfer through the building skin. Heating loads are calculated using:

$$q_{tot} = q_c + q_s + q_i$$

Cooling loads are calculated using:

$$q_{tot} = q_p + q_m + q_i + q_{CLTD} + q_r$$

There are several other concepts we have touched upon along the way. Mean radiant temperature is calculated based on the view angle and comparative temperatures. The temperature gradient inside of the wall is calculated based on comparative resistances and the overall temperature drop.

Finally, we have introduced a number of new terms, which may be the most important of all. Do you understand emissivity (ε), sensible heat, latent heat, specific heat (C_p), phase change, resistance and U value?

Take the quiz at the end of this lesson and when you are fairly confident about the information, proceed to the next lesson. Please make sure you go back to review anything you missed. Lessons Three through Five are sequential, and you should therefore understand one before proceeding to the next.

LESSON 3 QUIZ

1. Insolation is an example of
   A. radiation.
   B. conduction.
   C. convection.
   D. latent heat transfer.

2. Which of the following is dependent on orientation?
   A. Radiation
   B. Conduction
   C. Convection
   D. Latent heat transfer

3. The mean radiant temperature relates to
   A. air temperature.
   B. surrounding surface temperatures.
   C. body temperature.
   D. all of the above.

4. CLTD approximates the effects of which of the following?
   I.   Radiation
   II. Conduction
   III. Convection
   IV. Latent heat
   A. I and II   C. III and IV
   B. II and III   D. I and III

5. The resistance of a layer of construction in a wall relates to which of the following?
   I.   The temperature gradient within the wall
   II. The conduction through the wall
   III. The conductivity of the material composing the layer
   A. I only   C. III only
   B. II only   D. I, II, and III

6. A degree day is directly related to
   I.   the weather in an area over a period of time.
   II. a reference temperature.
   III. a design day.
   A. III only   C. I and II
   B. II and III   D. I and III

7. Infiltration is calculated using which method(s)?
   I.   Design day method
   II. Air change method
   III. Crack method
   A. I only   C. III only
   B. II only   D. II and III

8. A calculation of heat loss includes
   A. q_c, q_{CLTD}, q_p, and q_m.
   B. q_c, q_{CLTD}, q_s, and q_l.
   C. q_c, q_s, and q_i.
   D. q_s, q_i, and q_r.

9. Using the *design day* temperatures in the q_c formula gives us information for which of the following?

    A. Equipment sizing

    B. Life cycle costing

    C. Latent heat values

    D. Yearly weather versus a reference temperature

10. Which of the following is contributed to the environment by a tea pot boiling on a stove?

    A. Sensible heat

    B. Convection

    C. Latent heat

    D. All of the above

COMFORT, CLIMATE, AND SOLAR DESIGN

INTRODUCTION

We learned about the basic thermal processes in Lesson Three. We will now take that information and add to it certain ideas about what conditions are comfortable thermally, and some tools used to study and express different climates and human comfort ranges. After studying various climates, we will look at the shape traditional buildings take in response to those climates, and why. We will learn the basics of solar position and orientation. Finally, we will distill some old ideas out of the vernacular design, and add some new ideas and developments, which will result in what is generally called passive solar design.

HUMAN COMFORT RANGES AND ZONES

An Englishman is not comfortable in the same range of temperatures as a Tahitian. Human comfort ranges vary depending on culture, recent exposure and even health and age. However, the same factors always come into play. Let's begin by studying what those factors are.

Metabolism

In our discussion of cooling loads we considered the fact that humans generate somewhere between 450 and 2,500 Btuh. That heat must be dissipated from the human body by the same heat transfer mechanisms that remove heat from buildings. If that heat is not removed, body temperature begins to rise, which is uncomfortable at first, then unhealthy, and finally, even fatal.

In colder temperature ranges, radiation, conduction, evaporation, and convection remove heat at a steady and rapid rate. We are warmer than our surroundings and the heat loss is much greater than the gain. In order to limit the heat loss, the body tends to close down the pores, which reduces evaporation and latent heat transfer. It raises goose bumps and contracts the capillaries, reducing blood flow near the skin, which reduces skin surface temperature (thus reducing radiative loss and conductive loss). If necessary, we put on additional layers of clothing, providing insulation and trapping convection.

As the temperature rises, the heat loss due to conduction, convection, and radiation slows down, and the body begins to adjust. First the pores open up, the blood flow near the surface increases, and finally, we begin to sweat. The moisture evaporates, using up the latent heat of evaporation and removing it from the body. When the temperature of the environment reaches 98.6°F, all heat loss from conduction,

convection, and radiation ceases. When the temperature rises any further, the direction of the heat flow reverses, since the body tries to remain at 98.6°F. At this point, the only mechanism still removing heat from the body is the evaporation of sweat and of moisture from the lungs and sinuses. This is why humidity is such a key factor. If the air around us is already at 90 percent relative humidity, the sweat will hardly evaporate. To understand the relationship of all of these factors, we will use a single chart called the *psychrometric chart*. This chart is immensely useful and works for just about everything engineers like to calculate, far beyond simple questions of human comfort.

The Psychrometric Chart

The psychrometric chart is a graph that shows the air at different temperatures and different humidities. You can also graph the total amount of energy stored in the air (sensible heat + latent heat) on the same chart, and the term typically used for that combined storage is called *enthalpy*. To cool and simultaneously dehumidify air requires removing both forms of stored heat, so we can just look at the change in enthalpy.

Lines representing the dry bulb temperature (and constant stored sensible heat) run vertically on the chart. Lines of wet bulb temperature (and constant stored enthalpy) run diagonally from the lower right to the upper left. The wet bulb temperature is the temperature measured using a thermometer with a wet sock on the bulb so that the rate of evaporation is taken into account. Dry air results in a large *wet bulb depression,* the difference between the dry bulb and the wet bulb temperatures. The thermometer with a sock is called a *psychrometer.* If it is swung around in the air manually to get air movement, it is called a *sling psychrometer.*

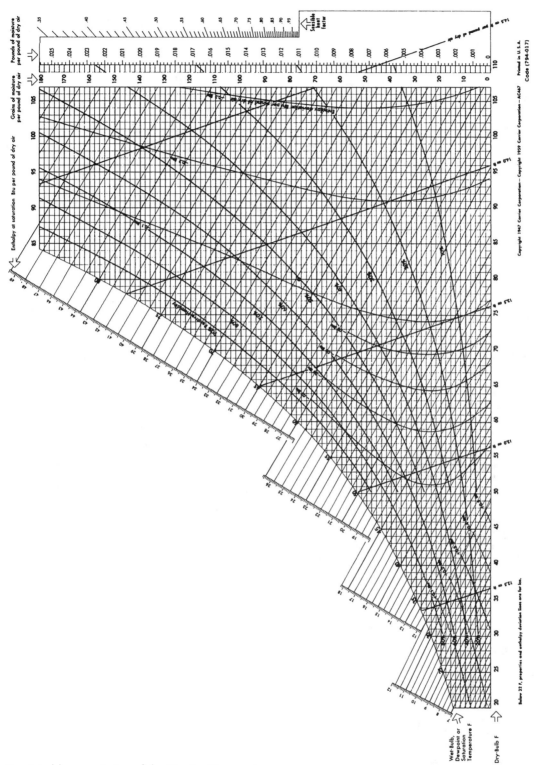
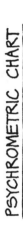

PSYCHROMETRIC CHART

Reprinted by permission of the Carrier Corporation.

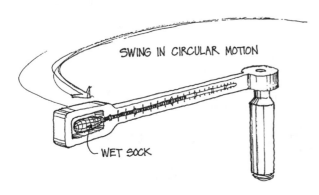

SWING IN CIRCULAR MOTION

WET SOCK

SLING PSYCHROMETER

The amount of water contained by the air is represented by the horizontal lines, either in *grains of moisture per pound of air*, or in *pounds of moisture per pound of air*, or in *pounds of moisture per 1,000 cubic feet of air* (even engineers don't agree on everything). *Relative humidity* is shown by curved lines that run from lower left to upper right. Note that a constant amount of water in the air does *not* represent a constant relative humidity (RH). Relative humidity is defined as the *percentage* of complete *saturation* (how much water is in the air at a given temperature compared to how much the air could hold at that temperature). Air can hold much more water when it is warm than when it is cold. So 0.010 lbs. of H_2O per lb. of air represents 90 percent RH at 60°F but only 20 percent RH at 105°F. This is why there is condensation on the outside of a cold glass of water or iced tea. As the warm, humid air gets cooled by the glass, it can't hold as much moisture, so some of it condenses out. This is also why we *always* put a vapor barrier on the *warm* side of the insulation in a wall. The temperature drop across the insulation will result in the air being able to carry less moisture. If we allowed moist air from the warm side (i.e., from inside the house) to penetrate into the insulation, the moisture would condense out *within* the insula-

tion, reducing its resistance, and in many cases causing building materials to deteriorate.

We can determine roughly what combinations of air temperature and relative humidity are comfortable for most people. We can outline that range of combinations on the psychrometric chart. That range of combinations is called the *comfort range* or *comfort zone*. In the United States, for people doing calm or sedentary work, and who are lightly clothed, it varies from 65°F to about 78°F and from 25 percent RH to about 75 percent RH. The higher RH values require slightly lower temperatures. For example, 75 percent RH is only pleasant up to 73°F. But at 25 percent RH, 78°F is just fine.

These comfort ranges are affected by factors which do not show on the psychrometric chart, for example, mean radiant temperature. We therefore use the term *effective temperature* for a combination of the ambient air temperature (dry bulb) and the MRT. If the MRT is high, then the comfort zone shifts to lower ambient air temperatures to make up for the difference, and when the MRT is low the comfort zone shifts to higher air temperatures.

Similarly, the local velocity of the air has an effect. When the air is moving, it carries away heat and moisture more rapidly. Thus, moving air shifts the comfort zone to slightly higher temperatures. If you sit in front of a fan, you can tolerate the heat better, even though you have not changed the temperature or the relative humidity.

SOLAR DESIGN

There are two environmental factors to which the design must respond: the sun and the climate. We will discuss the sun in sufficient detail for you to understand the basic concepts, and to handle any questions that the exam is

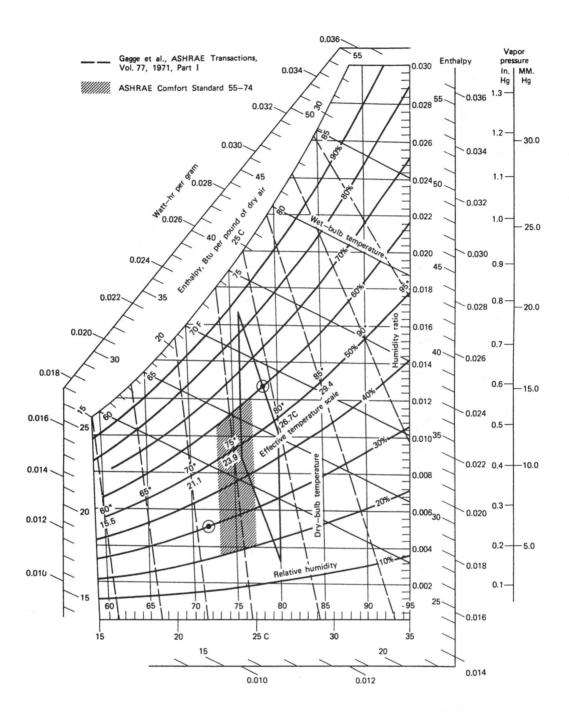

COMFORT ZONE

Reprinted from *Handbook of Fundamentals* by permission of ASHRAE.

likely to pose. This subject is worth a great deal more attention, and you are encouraged to study it further, because it is part of the global issue of energy. How you, as an architect, respond to the sun and other design factors will affect energy usage for decades to come. The following section should only be a beginning. For further study, check the references in the bibliography.

Solar Angles

The earth orbits the sun, and also rotates about its own axis, giving us day and night. The axis about which the earth rotates is not perpendicular to the plane of the orbit, but tilted at 23.5°. This means that part of the year the North Pole faces away from the sun, and part of the year it leans toward the sun. The seasons are caused by this changing *tilt* of the earth with respect to the sun, not by the changing *distance* of the earth from the sun. When it is winter in the Northern Hemisphere, it is summer in the Southern Hemisphere. The tilt of the North Pole in relation to the position of the sun is called the *declination* angle (δ). On December 21

(*winter solstice*), the declination is $-23.5°$, and on June 21 *(summer solstice)* it is $+23.5°$. In spring and fall it passes through a midway point ($\delta = 0°$) on March 21 and September 21 *(the equinoxes)*.

The declination tells us about the sun's *seasonal* relationship to the earth. To describe the sun's position relative to a given site at a given time of day, we use two angles. The first angle, the *altitude* angle (ALT, α or h) describes the height of the sun in the sky. The second angle, the *azimuth* angle (AZ, a_s, or bearing) describes the compass orientation of the sun. The altitude angle is always measured straight up from the ground. The azimuth angle, as commonly used in architecture, is the sun's position either east or west from due south. For example, if the sun is due south, AZ = 0°; if the sun is due east, AZ = 90° east of south; and if the sun is due west, AZ = 90° west of south. Another convention measures from due north in a clockwise manner. Thus due east results in AZ = 90°, south results in AZ = 180°, and west results in AZ = 270°. Computer programs, even in architecture, tend to use this convention.

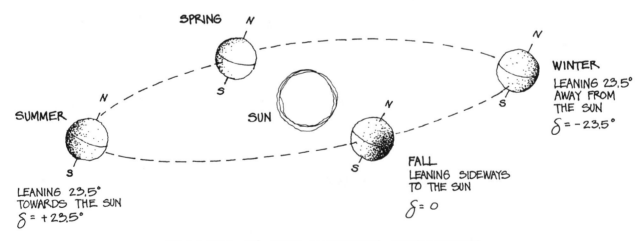

SEASONS IN THE NORTHERN HEMISPHERE
CAUSED BY CHANGING DECLINATION

Consider the sun's behavior in the Northern Hemisphere. In the winter the sun rises south of due east, arcs low through the sky and sets south of due west. In the summer it rises earlier and to the north of due east, arcs high through the sky and sets later, north of due west. At noon the sun is always due south, no matter what the season (north of the tropics). If you notice that sometimes this is not the case, it is because of the difference between legislated time and *sidereal time* (real or solar time). They are the same only at the center of a time zone, but elsewhere, sidereal time is either earlier or later than legislated time. In addition, we have daylight-saving time for six months of every year, which advances the "legal" time one hour.

The first conclusion we can draw from the seasonal variation in the sun's path is that there is a lot of sun on south-facing facades during the winter, and not much sun anywhere else. In the summer, on the other hand, there's a great deal of sun on the east facade, the roof, *horizontal skylights*, and on the west facade. This means that southern exposures should be maximized for winter sun, especially if there is an overhang, while eastern and western exposures and horizontal skylights should probably be minimized to reduce summer sun.

Indeed, in temperate and cold climates, this is the basic passive solar design strategy. We want maximum glass on the south facade, but the sunlight that comes in through the glass in the winter should be stored without changing the temperature too rapidly, which would cause overheating. For this reason, there should be sufficient thermal mass, which could be provided in the form of a concrete floor with a conductive finish such as slate or terrazzo. Mass could also be provided by a concrete block wall or other heavy building material. Special solar design prototypes will be discussed later.

Overhangs

South-facing windows should be protected by an overhang that allows the low winter sun to pass under the overhang and in through the window, while the high summer sun is cut off. Thus, the perfect overhang is one that is calculated to admit the winter sun and block the summer sun. This can be done by placing the overhang higher up the wall (at a greater

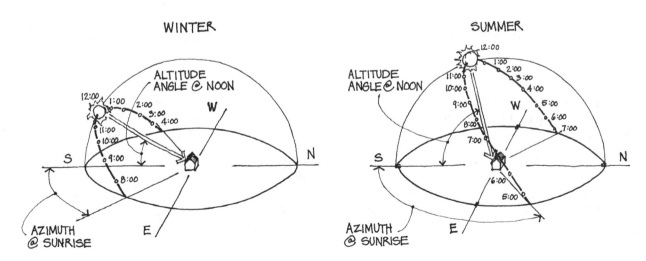

SUN'S PATH ACROSS THE SKY AT 40° LATITUDE

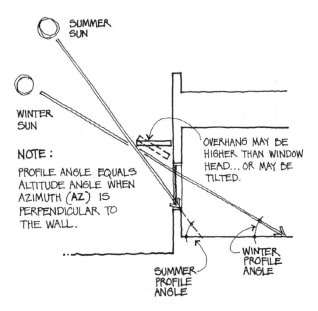

OVERHANG AND PROFILE ANGLE

distance above the window head) and increasing the projection proportionally, or by tilting the overhang to match the incoming winter ALT angle, as shown above.

The angle of the *shadow* line is called the *profile angle* (Φ) and coincides exactly with the altitude angle when the sun is directly facing the wall (perpendicular azimuth). At other times of the day, the profile angle is determined by the interrelationship between azimuth and altitude angle, which varies by season.

Fins

In temperate or hot climates, we can afford very little glass on the east or west facade, especially since the west facade gets heat in the afternoon, when the outside air temperature is highest. *Horizontal* glass should be minimized as well. Clerestories or even lanterns or sawtooth roofs should be used instead of horizontal skylights. In cold climates, the sawtooth can face south (although we have to be careful about glare on the space below), and in warm climates we face the sawtooth north.

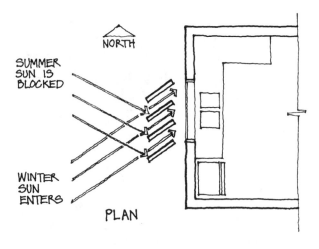

VERTICAL FINS ON WEST FACADE

East and *west* facades should be protected with *vertical* fins or a combination of horizontal and vertical fins. If we turn the fins to point slightly south, then the winter sun will be admitted (the sun sets south of west in winter) but the summer sun will be greatly reduced (the sun sets north of west in summer).

Solar Plot

The design tools developed to handle all of this are called the *solar plot* and the *shadow mask*. The solar plot is the path of the sun plotted onto a grid (circular or rectangular), and the shadow mask is the representation of the shading devices plotted onto the same grid, so that they may be compared.

The circular graph comes from drawing a hemisphere over the site. The solar plot comes from the sun's path on that hemisphere, and the shadow mask comes from masking all of the angles that are obscured by the shading device (or trees, other buildings, surrounding terrain, etc.). The radial lines are azimuth lines and all of the concentric circles are altitude lines. While we can plot the path of the sun for any given day, typical diagrams plot one day for each month. This results in seven curved lines, one

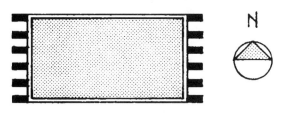

VERTICAL FINS ARE
MOST EFFECTIVE IN EAST
AND WEST EXPOSURES

each for December and June, and five for months with identical paths (e.g., January and November, February and October, March and September, etc.). Only December and June are unique.

The same information can be drawn on a rectangular graph, in which case the vertical lines are the azimuth values and the horizontal lines are the altitude angles. The shading mask is very much like a wraparound photo of all of the objects that would obscure the sun and cast a shadow on the point being considered.

The shadow mask allows the designer to check what time of day or year the sun will fall on the window, and when it will not. If a tree or a shading device covers a certain set of angles on the plot, then it would result in the window being shaded for those sun angles. The designer may add a tree, or an overhang or a fin. Conversely, if another building causes too much shadow at the wrong times, he/she may move the window.

Solar Intensities

There is a secondary result of the solar position. When you shine a flashlight against a wall you get a small round spot of high intensity if the flashlight is held perpendicular to the wall. If you aim the flashlight at an angle, you get a larger elliptical area of lower intensity.

Similarly, the sun's intensity per square foot varies depending on the angle between the wall and the *solar vector* (a line drawn to the sun's position). The highest intensity occurs when a hypothetical receiving surface is perpendicular to the solar vector, and that intensity is called the *direct normal intensity* (I_{dn}).

This direct normal intensity varies with the time of day. In the early morning, because the sun is at a shallow angle, the sun has to pass through a lot more atmosphere. This reduces the intensity available. The combined effect of varying direct normal intensity, and of varying angular relationship between wall and sun, results in a very large variation in the *solar intensity* (I_s or S_g) on a given wall orientation for a given time of day. If you remember the discussion of radiant gain (q_r) and solar intensity (S_g), you will remember that S_g varies depending on orientation and time of day. It is the result of the combined effects of the altitude angle, the azimuth, and I_{dn}.

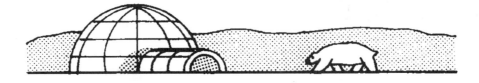

EXTREMELY COLD CLIMATE....
... COMPACT ARCHITECTURAL FORM

CLIMATE

There are many classifications for climate. The best way to study a climate is to plot it on the psychrometric chart. Let us look at one of the four prototypical climates on a psychrometric chart. Although there are numerous climate zones for calculating energy code compliance, discussing four prototypes is sufficient for our purpose.

Cold

The cold prototype is actually rare within the United States; Alaska and the North Central plains states probably qualify. Any time you see a car from Montana, look for a plug sticking out of the grill. It is there so that the engine block can be plugged in overnight to keep it from freezing.

If we plot the daily temperature cycle for an average day of each month in Billings, Montana, we see that most of the year falls to the left of the comfort zone, meaning that it is too cold. However, certain times of day during the summer months of June, July, and August fall within the comfort zone. If we could capture daytime temperatures and seal the building at night, we would be in great shape. Also note that the daily loop is almost always the same shape and angle. The temperature and the relative humidity are inversely proportional to each other (high temperature results in low RH, low temperature results in high RH). The peak temperature is around 3:00 or 4:00 in the afternoon, and the peak relative humidity is coincident with the minimum temperature, at 4:00 in the morning. It is not that a lot of

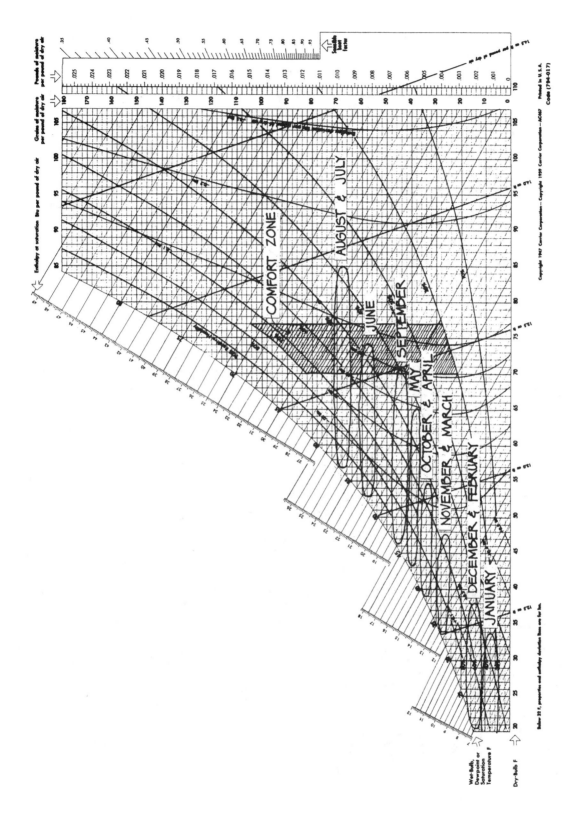

COLD CLIMATE PLOT
(BILLINGS, MONTANA)

moisture has been added to the air, but rather that the same moisture content represents a higher percentage of saturation at the lower temperatures. If the air temperature drops below the temperature at which saturation occurs, we have 100 percent RH, and the moisture will condense out. This either means rain, or sometimes, just dew. Thus this temperature (for a given moisture content) is called the *dew point temperature.*

The best solutions for a cold climate are building forms that minimize the exposed surface area (remember, $q_c = U \, A \, \Delta T$). The objective is to enclose as much volume as possible within the minimum envelope. The best forms tend to be the simplest; for example, the cube or the hemisphere. An igloo is really an ideal form given the building materials and climate. Most vernacular designs in colder climates are cubical buildings,

typically two stories with a big sloping roof. Many cultures have farmhouses that include the barn under the house on the ground floor, and the living space on the upper floors, enclosing all the volume in a single skin. The same basic farmhouse form can be found in Switzerland and Nepal. New England homes had very blocky houses with most of the glass on the south facade. Storage rooms, which weren't terribly important, and the kitchen, which created its own heat, ended up on the north. Evergreen trees were planted in a row or in a large group to the north or northwest of the building, to block the prevailing winter winds. A row of trees will shield objects downwind for a distance of three to five times the height of the row.

The classic salt shaker is a house with a two story south facade, and a long sloping roof to the north, resulting in a single story north facade.

SALT SHAKER

If the builder was confident in his ability to build a firesafe chimney, the fireplace was placed in the middle of the building, adding internal thermal mass. If not, it was on the west end. The entry had an internal or external *vestibule* (now known as an *airlock*) that allowed people to enter or leave without losing all of the heat from inside. The house performed extremely well for the building technology available at the time.

Temperate

Most of the United States has a temperate climate. The winters are too cold and the summers are too hot. The climate plots are below the comfort zone in winter, above the comfort zone in summer, and shift through the zone in spring and fall. If we use air movement, solar input to boost MRT, and proper insulation, we can solve both kinds of problems. The best way to

do that is to ascertain from the plot what months require what action, and then superimpose those months on a solar path diagram to see what solar angles should be admitted and when shading is necessary.

The resulting building shape tends to be a modified version of the cold climate building. It has been stretched to the east and west, making the south facade longer (remember the discussion of solar paths in winter and summer). There are often porches to the south, or awnings on the windows. This results in an increase in winter solar gain, without any increase in summer solar gain. We sometimes find a large deciduous tree on the south or west, which provides shade in summer and which loses its leaves in winter. To the north we again find evergreens, if there is any northerly winter wind.

TEMPERATE FARM HOUSE

TROPICAL HOUSE

Hot Humid

The climate plot for a hot humid climate, such as that of Houston, is perhaps the most troublesome. It is out of the comfort zone for much of the time simply because of the humidity level. Furthermore, some of the strategies that designers use for other climates don't help here. Until there was air conditioning, insulation was of no help, and we certainly do not want solar gain. The best that could be done was to allow whatever air circulation there was to dissipate heat as rapidly as possible.

The buildings loosen up even further from the temperate prototype. The kitchen detaches, or on more elaborate buildings, there are two kitchens: one for summer and one for winter. Often there are exterior passageways of one sort or another, balconies or breezeways. In really tropical climates, the walls are sometimes omitted, or reduced to lightweight privacy shields. The house is sometimes elevated on poles to allow air flow underneath. The roof is in two separated layers or completely open at the gable so that warm air rises up and out of the building. Some Southern houses actually use this method to empty out the air through the central stairwell. Using convection to suck fresh air through the building in this manner is called *thermosiphoning*.

The brush is often cleared from window areas to allow the best possible ventilation, and palm trees (if available) are planted. The palm is the closest thing to a natural parasol: there is no impedance to breezes underneath, but still some shade when several trees are grouped together.

Hot Arid

A hot arid climate plot typically shows the greatest daily variations. Hot arid climates, such as that of Phoenix, tend to have large diurnal (day to night) temperature swings. This is because places with an arid climate usually have a clear sky. At night, there are large radiation losses to the sky, both from the building and the environment in general. The large temperature swing

COURTYARD HOUSE

gives us a major opportunity, because if either extreme is within the comfort zone, it can be captured and stored by the building, resulting in a pleasant environment.

In addition, we know that heat can be "used up" by the evaporation of water. If water is available, the air is dry enough so that evaporation within the building compound can bring the temperature down significantly.

The vernacular design for such areas takes advantage of those two factors. It is built of high thermal mass materials, such as adobe. The adobe stores the heat from the day for the following night and stores the coolness of the night for the following day.

The form of the house is the ultimate extension of the progression from tight cube to loose elements: it wraps around the outside environment and pulls it in. The Spanish or Moorish atrium or courtyard house contains a piece of the outside environment in its center, which it modifies by evaporating moisture into the shaded area by the use of a fountain or plantings. The house is open to the courtyard, so the cooled and humidified air enters the house. The only windows on the outer walls are small and occur high on the

wall, allowing the heated air from inside to escape, but forming a "pool" of cooler air, which does not rise. There are psychological factors inherent in this kind of building, as well. The sound of the fountain and the quiet courtyard produce a cooling and calming environment, even if they don't bring about much change in temperature.

PASSIVE SOLAR DESIGN PROTOTYPES

Passive solar design uses the sun to heat the building without any moving parts or pumps. In a sense, this definition is accurate but oversimplified. Passive design also includes buildings that do *not* overheat because they are correctly designed. In fact, it should include all buildings, since all buildings (whether "solar" or not) should take proper orientation and materials into account. The strict definition of a *passive system* is one in which the collector device and the storage device are one and the same (i.e., the structure itself). *Active systems* are those in which the collector and storage are separate (i.e., a collector on the roof and a storage tank in the basement). Most buildings are somewhere in between.

The first principle of good passive design is to do whatever the good vernacular design solutions did. This results in appropriate regional architecture, usually responds to the context, and represents an optimum use of form and materials.

Direct Gain Space

There are some special strategies that have been developed, and some generic names have been given to old strategies. For example, a room in which the structure and thermal mass are in direct sunlight is called a *direct gain space*. This usually means that uncovered, high mass floors are placed in the southern rooms, so that they absorb the sunlight directly. Concrete, stone, terrazzo, tile and other materials that conduct and store heat are ideal for this purpose.

Mass Wall

The first extension of this strategy is to place specially thickened walls directly in the sunlight (typically on the south facade). These walls are often behind a large window, or sometimes an entire glass skin, and are generically called *mass*

walls. They store the incoming solar energy without increasing temperature rapidly, and release the heat when needed without dropping temperature rapidly. They often store a great deal of heat. Mass walls also delay the arrival of the heat on the inside, and may be tuned in thickness to match the expected temperature cycles. There are two special types of mass walls: Trombe walls and water walls.

Trombe Wall

The most famous type of mass wall is the *Trombe wall*, named after a Frenchman, Felix Trombe, who popularized the technique. The Trombe wall adds a convective loop to the system, by trapping a layer of air between the wall and the external glass skin. When the air heats up, it rises. There is a one-way vent at the top of the wall that lets the warm air into the room. There is another one-way vent at the base of the wall that lets the coldest air in the room, which is along the floor, into the space between the glass and the wall (which, in turn, is heated, repeating the cycle). This is an

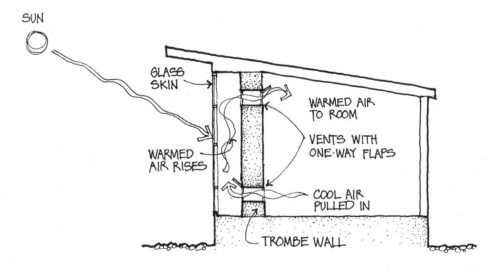

SUN

GLASS SKIN

WARMED AIR RISES

WARMED AIR TO ROOM

VENTS WITH ONE-WAY FLAPS

COOL AIR PULLED IN

TROMBE WALL

TROMBE WALL

example of a type of thermosiphoning. The wall literally stirs the room air when the sun shines and shuts itself off when the sun does not shine.

Water Wall

The second type of mass wall is the *water wall,* which consists of a tank or a collection of large vertical tubes, filled with water and placed next to the window, inside the building. The tubes and the water may be clear or colored. It is different from other thermal mass materials in that it allows some of the light through the wall into the room. Water has an extremely high specific heat, storing about 5 times as much heat per degree change per pound as concrete does (or about two to three times as much per cubic foot). Thus, it mitigates temperature fluctuation within the room. (See discussion of C_p in Lesson Three.)

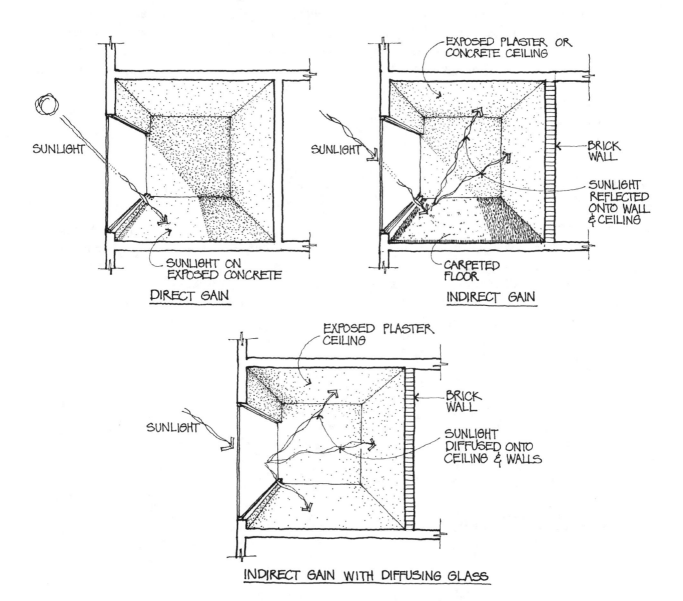

DIRECT GAIN

INDIRECT GAIN

INDIRECT GAIN WITH DIFFUSING GLASS

Indirect Gain

An *indirect gain* space is similar to a direct gain space, in that the usual structural and finish materials have high thermal mass and can absorb and store a great deal of heat without much temperature fluctuation. The difference is that in the indirect gain space, the mass is not directly in the path of the sunlight, but in a shaded area of the room. The thermal mass is heated indirectly, either by reflected sunlight or by the warm air in the room. This strategy can be improved immensely by using a diffusing glass in the window, which redirects the incoming sunlight to hit much of the mass directly. Otherwise, an indirect gain space requires about four times as much thermal mass as a direct gain space, to have the same effect.

Greenhouse

Perhaps the most common simple application of solar design is the addition of a *greenhouse* onto a living space. You remember the definition of the greenhouse effect given in Lesson Three. Typically, a fan is connected to a temperature sensing device and moves the air from the greenhouse into the space when the desired temperature differentials are attained. There should also be an exhaust fan for use when the greenhouse overheats.

Super Insulated

Another old technique that has been revived and perfected is the use of very large amounts of insulation, and most importantly, sealing the construction very carefully. Super insulation, consisting of R-20 walls and R-30 roofs, is used; the seams on vapor barriers are carefully taped and gaps between window frames and wall are foam filled. There are no pipes or conduits in exterior walls. The electrical wiring, switches, and sockets are typically mounted on the room *surface* of the wall in a conduit or chase rather than penetrating the wall.

Double Envelope

Double envelope construction consists of a building within a building. The outer shell utilizes passive design, with lots of south-facing glass. The inner shell also involves passive design, with a backup heating system. The inner shell is maintained at the intended temperature, and the outer shell provides a mild and protected climate for the inner shell. The east and west facades often consist of a single superinsulated shell. The result is that

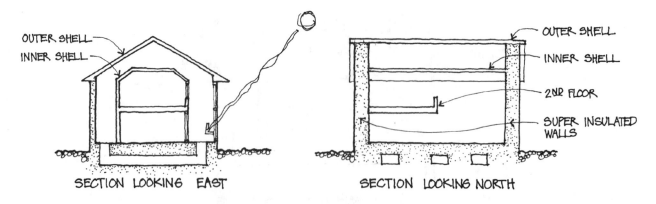

DOUBLE ENVELOPE CONSTRUCTION

a north-south cross section looks like one building inside another, while an east-west cross section looks like a single building with two roof layers. It was originally believed that there would be a convective loop around the building when the sun shone on the south side. However, experiments indicate that some stagnation occurs and that this effect is not as predominant as originally surmised. There does seem to be a connection between the space between the inner and outer shells and the temperature of the ground under the building, resulting in fairly good performance anyway.

Earth Sheltered

Perhaps the most interesting development is the trend toward "earth sheltered" buildings, which may be partially sunken, "snuggled into the hillside," bermed, or totally underground. The greatest thermal benefit in all of these cases is *not* the *insulation* provided by the earth, but rather its tremendous *thermal mass*, with the resulting storage and averaging effects. If you dig deeply enough into the ground, you will come to a layer that never experiences freezing temperatures (the frost line). In fact, we can draw temperature profiles, in which the surface of the ground may be at air temperature, but just

a few inches lower, the temperature is much milder, and at a depth of 20 feet, there is almost no variation from the year round average. Year round average temperatures are much easier to deal with than the extreme air temperature swings we get in most climates. With earth sheltered buildings, there are increased structural costs, and waterproofing is critical, but there are a number of benefits as well: increased security, durability, and privacy, and decreased maintenance, since no short term painting and reroofing are necessary. In addition, more ground surface becomes usable as outside space.

Other Passive Techniques

There are other techniques that are more or less "passive" in nature. In climates with high daytime temperatures and low nighttime temperatures, ventilating the building at night to cool it and closing it during the day can keep the building as much as 20° cooler than the daytime peak. This is called *nighttime flushing*.

The *roof pond* uses sliding insulation panels over a pond or bag of water on the roof. During the summer daytime, the panels are closed and the water absorbs heat radiated from the house below. At night, the panels are opened and the

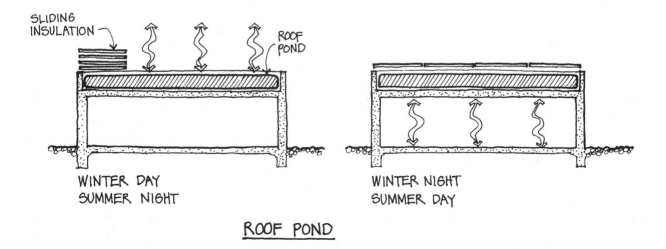

ROOF POND

heat is reradiated to the nighttime sky. This requires comparatively clear climates, since cloud cover reflects ground radiation back down. During the winter, the system may be used in reverse if needed. The panels are left open all day long, exposing the water to the sun, and closed at night to insulate the heat that was gained. This only works, again, where the air is clear and the temperatures mild enough so that the water heats up. In such instances, the water is most often placed in large, black plastic bags, increasing both absorptivity and emissivity, and the insulation panels are foil backed on both surfaces. Radiation is the key thermal process employed.

In climates with snowy winters and hot, humid summers, there have been buildings built in conjunction with an insulated pond, often covered by an insulated dome. All winter long, a snow machine sprays a fine mist of water into the pond, while the dome is opened to the outside air. An enormous pile of heavy snow is formed where the pond used to be. In the spring, the dome is closed off, and the snow/ice begins to melt. The resulting water provides an excellent source of chilled water for air conditioning systems. These projects are economically feasible primarily for rural or suburban campuses, where there is sufficient low cost land for the ice pond.

Energy from the sun or the wind, or even from burning wood, is know as energy from *renewable sources*. Such energy may be utilized for either passive or active systems. Let us look at some of the active systems.

ACTIVE SOLAR SYSTEMS

Active solar systems are typically used to do one of four things: heat water, heat the building, cool the building, or generate electricity. All active systems have some elements in common. These include flat plate or focusing collectors, and rock bed, fluid container, or battery storage devices. In some cases, the existing utility network is actually used as the storage device.

A *flat plate collector* is exactly what the name implies, a flat surface tilted at approximately the right ALT and AZ angles to receive most of the sun's rays as directly as possible. Such collectors usually continue to function even when the sun is not exactly perpendicular to the surface, and also on overcast days when the solar rays are scattered and not parallel.

A *focusing* collector consists of either a parabolic trough or a parabolic dish, or an arrangement of lenses, which focuses incoming light onto a tube or a point. It generates much higher energy densities and much higher temperatures than a flat plate collector. However, it also has to be aimed much more carefully. If the collector is not properly aligned with the sun, then the sun's rays focus on the wrong point altogether. If the sky is cloudy and the solar rays are not parallel, then focusing collectors often don't work very well no matter where they are pointed. This means that most focusing collectors are mounted on gimbals or motorized devices that allow them to be constantly adjusted throughout the day. They are most effective in clear climates.

Domestic Hot Water

The most effective use of active solar systems at this time is to heat water to be used in place of domestic hot water or hot water used in industrial processes. There are several different types of systems, which may be either open or closed loop.

A flat plate collector system uses a collector plate composed of tubes that either have flat fins or are attached to a large metal collector surface. The surface is insulated heavily on the back and

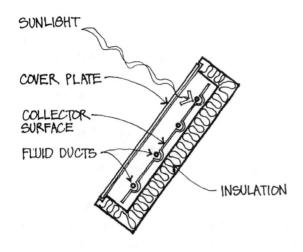

FLAT PLATE COLLECTOR

The black tubes are copper or steel pipe with the water running through them.

Some systems use a bent *Fresnel lens* to focus the incoming sunlight. A Fresnel lens uses less material than a normal lens. The surfaces still have the same orientation and curvature, only they are stepped, rather than continuous. The most common Fresnel lens is an automobile headlight, the inside of which is ridged to help focus and direct the light without using a thick lens in front of each lamp.

sides, and has a cover plate, typically of a very high transmissivity glass or plastic on the front.

A focusing collector system utilizes a parabolic trough with an interior surface of reflective aluminized mylar and a tube, clear or black, running through the focal line of the trough. The clear tubes are usually the size of a fluorescent tube and filled with a colored collector fluid.

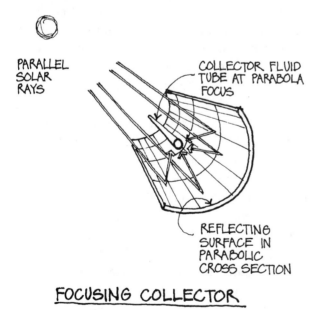

FOCUSING COLLECTOR

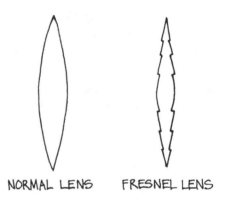

NORMAL LENS   FRESNEL LENS

FRESNEL LENS

Open loop simply means that the fluid going through the loop is the fluid that will be consumed. For example, the water in the loop will be used for domestic purposes such as cooking or washing.

Closed loop systems use one medium in the collector, which then runs through a water storage tank without being mixed with the water. Usually antifreeze (glycol) is run through the collector and then through a coil inside the water tank. This solves the problem of what to do with the fluid inside the collector overnight. If water were used inside the collector in a cold climate, it could freeze and burst all of the tubes in the collector as well as the connecting piping.

Drain down and *drain back* systems solve the freezing problem by emptying the collector when the temperature drops too low. The drain down system senses the temperature differential and opens valves to drain the fluid down into a reservoir. The drain back system is a fail-safe system in which the collector is only full while the pump is running. If the pump shuts off, everything in the collector drains back. The pump is set so that it only turns on when the temperature in the collector is higher than in the storage.

A *batch system* is nearly a passive system, and is simply a storage tank exposed to the sun. In many cases the tank is encased in an insulating box with a folding insulating panel over the glass face of the box. For this reason batch systems have sometimes been nicknamed *"bread-box"* systems.

A *thermosiphon system* is one in which the storage tank is higher than the collector and often immediately adjacent to it. The water circulates between the collector and the tank by convection rather than being pumped. The warmed water in the collector rises into the top of the tank. The coldest water, at the bottom of the tank, is siphoned off to the base of the collector. If the tank is outside the house, it must be well insulated.

Space Heating

Active solar collectors are used for space heating in two ways. A simple and safe way is actually halfway between an active and a passive system. This system is known as *air and rock bed storage*. The collector may be a large flat plate with air ducts rather than water ducts, or even a greenhouse type of space. The heated air is blown through ducting to a large bin full of coarse gravel, often underneath the house. The air passes between the chunks of gravel, warming them and storing the heat. Later, when the sun has set, the air may be blown through the rock bed in reverse, and then ducted to whatever portion of the house requires heating at the time.

The other method uses water as the medium and is very similar to heating the water for domestic hot water use. The only difference is that the water is piped either to a heat exchanger at the furnace, or more often, directly around the building to fan coil units, or to radiant surfaces or registers, or to baseboard heaters.

Absorption and Desiccant Cooling

It is not surprising that cooling with solar energy is much more difficult than heating. Two methods have been developed. For small scale applications, the *desiccant system* is being tested. This system uses the sun to bake all of the moisture out of a desiccant. Outside air is then brought in past the desiccant, which absorbs all of the moisture. The dried air may be passed into the building, or water may be sprayed through it, the evaporation causing the temperature to drop. The system requires two batches of desiccant, one which is being dried while the other is being used.

The *absorption refrigeration* cycle is a fluid version of the same principle. The sun is used to evaporate moisture out of a brine solution, typically *lithium bromide*, until it reaches saturation. The solution is then used to absorb water *vapor* from a second source of clear water, increasing the evaporation rate in that source, and thus *cooling* it. Water from that second source may then be used for cooling the building. More often, a second water line is run through a heat exchanger in the cooled pool.

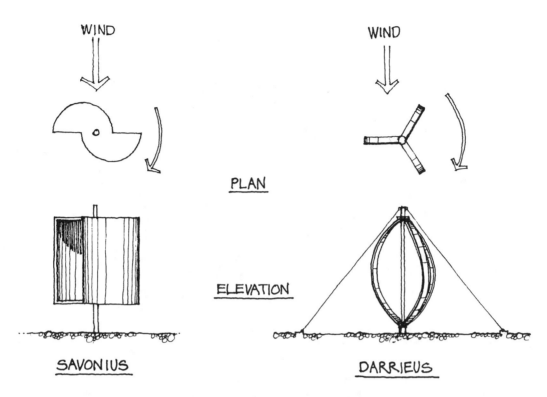

WIND

WIND

PLAN

ELEVATION

SAVONIUS

DARRIEUS

VERTICAL AXIS WIND TURBINES

Steam Generation

Solar heat in focusing collectors can not only warm water but can actually generate steam. The primary use for such installations is to generate electricity. On a small scale, this is quite safe, but not terribly efficient.

Photovoltaics

The most efficient method, with a nearly unlimited potential, is the direct generation of electricity from sunlight. This is called *photovoltaic* conversion, or more commonly, *solar cells*. Flat, very thin cells of a semiconductor made from silicon (sand) create an electrical charge (a difference in electrical potential) when they are exposed to light, which can be utilized simply by tapping opposite surfaces of the cell with small wires. This is equivalent to a direct current (DC) battery and can run lights and other simple electrical devices, or the electricity can be converted to alternating current (AC) and used with common house hold devices or sold back to the utility company. The most effective type of DC to AC conversion is called a synchronous inverter. Photovoltaic cells have been in use for some time, and are already cost effective where the alternative is to put up new power poles and run long power lines. The cost of photovoltaics is expected to continue dropping, much like the cost of calculators and other semiconductor devices over recent years.

The early cells were made of crystalline silicon, first cut from an ingot and later extruded in a ribbon. The most common cheap cell now is an *amorphous* silicon cell, often found on "solar" calculators. There is also some promise shown by gallium arsenide based cells.

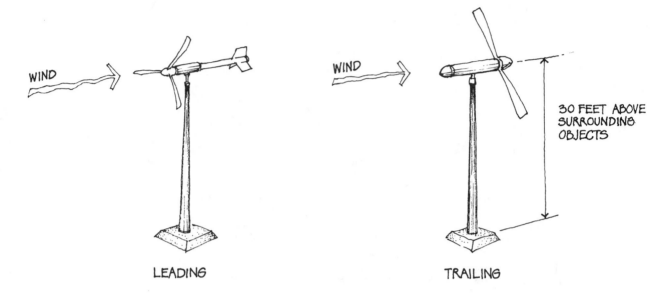

<u>HORIZONTAL AXIS WIND TURBINES</u>

Typical conversion efficiencies range from 10 percent to 13 percent. Efficiencies as high as 60 percent have been attained under laboratory conditions.

Wind Turbines

Another form of technology that has been with us for some time is the windmill. Wind turbines capable of providing for all the needs of a household can be purchased for under $20,000. They generate electricity through a generator or alternator, which can be used directly, or converted to AC and sold back to the utility. Large "wind-farms" in California, typically in a mountain pass, are now covered with turbines, and in New England, some consumers generate their own electricity using individual wind turbines. Any area with an average windspeed of 10 mph or more is a good candidate for a wind turbine, and a 13 mph average results in a steady profit from selling electricity to the utility company.

There are two basic types of wind turbine: the *vertical axis wind turbine (VAWT)* and the *horizontal axis wind turbine (HAWT)*. The VAWT is usually a Savonius or a Darrieus turbine. The *Savonius* is composed of two offset cups which spill into each other, much like two halves of a steel drum that has been sliced vertically and slid halfway apart. The *Darrieus* resembles a single eggbeater stuck in the ground. Both turbines work regardless of the direction of the wind. The Savonius is not as efficient, but is self starting. The Darrieus is very efficient, but will not start itself. However, it picks up speed nicely, once it is going.

The horizontal axis turbine is the more common type today, and is available in leading or trailing blade configuration. If the blades are upwind of the tower, it is called the leading configuration; if they are downwind of the tower, it is called trailing. Keeping the blades upwind (leading) requires either a tail, like a weather vane, or some arrangement of gears and a motor which is attached to a smaller wind direction sensing device. The downwind (trailing) configuration results in eccentric vibration, caused by the temporary release of pressure on the blade as it passes through the downwind "wind shadow" of the tower.

SUMMARY

We have discussed what conditions are comfortable thermally and some tools used to study and express different climates and human comfort ranges. We looked at four climates and at the shape traditional buildings take in response to those climates, and why. We reviewed the basics of solar position and orientation. Using that information, and what we knew of basic heat transfer methods, we examined different prototypical solar design strategies and the terms associated with them.

LESSON 4 QUIZ

1. If the MRT (mean radiant temperature) is low, which way does the comfort zone shift?

   I.   Toward higher ambient temperatures

   II. Toward lower ambient temperatures

   III. Toward lower relative humidities

   A. I only   C. III only

   B. II only   D. II and III

2. What is the definition of relative humidity?

   A. The amount of water in the air in lbs. of water per 1,000 cu.ft. of air

   B. The amount of water in the air in lbs. of water per pound of air

   C. The amount of water in the air in grains of water per pound of air

   D. The percentage of saturation at the current temperature

3. What is a sling psychrometer?

   A. A chart of the human comfort zone

   B. A chart of air temperatures and humidities

   C. A method of measuring wet bulb temperatures with a manual device

   D. A method of measuring the psychological effect of diurnal temperature variation

4. What is a Trombe wall?

   A. An unusually heavy mass wall

   B. A mass wall that uses water as the storage element

   C. A mass wall that causes a convective loop into the room behind it when it is warmed

   D. A mass wall that stores energy using phase change materials

5. A good design for a hot arid climate uses which of the following?

   I.   Thermal mass to damp out temperature swings

   II. Large openings to the outside to improve ventilation

   III. Evaporation to increase humidity

   A. I, II, and III   C. II and III

   B. I and II   D. I and III

6. The time of day affects which of the following factors?

   I.   The declination angle

   II. The altitude angle

   III. The azimuth angle

   IV. The intensity of the sun

   A. I and II   C. II, III, and IV

   B. I and III   D. I, II, III, and IV

7. Given the psychrometric chart on page 67, what is the wet bulb temperature if the dry bulb temperature is 60°F and the relative humidity is 70 percent?

   A. 50°F   C. 96°F

   B. 60°F   D. 104°F

8. Which of the following would NOT cause a change in enthalpy?

   A. An increase in the velocity of the air movement

   B. A decrease in relative humidity

   C. An increase in the dry bulb temperature

   D. An increase in the absolute moisture content

9. The major benefit to earth sheltered housing is derived from which of the following factors?

   A. The thermal storage capacity of the earth

   B. The insulative capacity of the earth

   C. The available moisture in the earth

   D. The weight of the earth

10. Which of the following statements is NOT true?

    A. Flat plate collectors will continue to function even when they are not pointed exactly at the sun.

    B. Focusing collectors can achieve higher temperatures than flat plate collectors.

    C. Flat plate collectors can be used for heating water or air.

    D. Focusing collectors are easier to mount than flat plate collectors.

MECHANICAL EQUIPMENT AND ENERGY CODES

INTRODUCTION

In the previous lessons we discussed how to reduce heating and cooling loads on buildings by being responsive to the climate and by using solar design, and we also described how to calculate thermal loads. In this lesson we will discuss the concepts and the mechanical equip-ment for heating, ventilating, and air conditioning (HVAC) a building in response to those loads.

Generally, this equipment consists of two basic elements. The first, called the *plant*, creates the warm or cool water or air, usually in the mechanical equipment room. The second is the *distribution* mechanism or *system*, which then delivers that heated or cooled air or water to the necessary rooms or areas in the building, called *zones*.

PLANT TYPES

The scale of a plant may vary, from a room air conditioner, through the equipment in a mechanical room in a building, to a steam plant used to heat an entire campus. Various plants may be used with various distribution systems.

Boilers and Chillers

The early plants were primarily for heating only and consisted of sources of hot water or steam. As air conditioning developed, plants became sources of chilled water or cooled air as well.

Water was heated by lighting a fire under a tank or heat exchanger tube, called a *boiler*. There had to be a separate exhaust *flue* to vent the byproducts of combustion from the building.

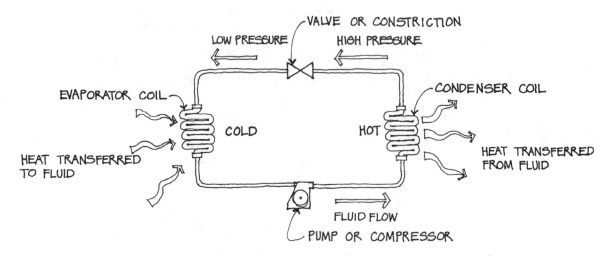

REFRIGERATION CYCLE

Today, in energy efficient designs, the air used for combustion is sometimes brought in directly from outside, instead of wasting the warmed and humidified building air. This is called *external combustion air.*

The modern *forced air furnace* duplicates the boiler in residential applications. Air from within the home is brought through a manifold inside a larger combustion chamber. Oil, natural gas, or propane is burned inside the chamber, warming the manifold, which warms the air inside. The combustion air is vented through a flue. Historically, convection moved the warmed *supply air* from the manifold up into the residence (this was called *gravity* feed) but this arrangement required the furnace to be located in the basement, and also didn't move the air fast enough. As a result, a fan was introduced, forcing the air returned from the home (*return air*) through the manifold and into the ducts no matter what the height relationships were. If the flow is downward through the furnace (reversing the convection) it is called a *downdraft* furnace. If the furnace is reduced to approximately five feet in height so that it can fit in an overgrown closet or even an attic it is called a *lowboy.*

Refrigeration Cycle

Modern air conditioning relies primarily on the refrigeration cycle. The refrigeration cycle uses a special fluid, *Freon* (which is actually a family of several chlorofluorocarbon or CFC gases), circulated in a closed loop. The *pressure* in the loop is varied using a pump and a constricted section of tubing or a valve, causing changes in temperature and evaporation and condensation.

The pump increases the pressure of the fluid, forcing the Freon to condense, thus releasing the latent heat of evaporation. This part of the loop is called the *condenser.* After the Freon passes through the condenser, it passes through an expansion valve, or simply a constriction in the tube, which results in a pressure drop on the downstream side. This drop in the pressure allows the liquid to evaporate, absorbing the latent heat of evaporation from its surroundings. This is called the *evaporator.* Because of the extreme pressure differences, the condensation goes on at a very high temperature and loses heat to its surroundings. The evaporation goes on at a very low temperature, and absorbs heat from its surroundings. This means we can move heat from a lower temperature to a

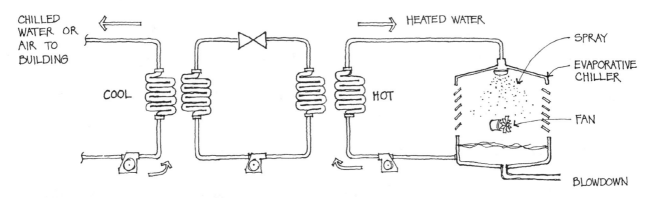

CHILLED WATER OR AIR TO BUILDING

HEATED WATER

SPRAY

EVAPORATIVE CHILLER

COOL

HOT

FAN

BLOWDOWN

REFRIGERATION IN BUILDINGS

higher temperature, which would otherwise be impossible.

Both the evaporator and the condenser are usually heat exchanger coils that heat or chill another fluid. On the condenser side the coil often transfers this heat into water being circulated through an *evaporative chiller* outside the building that dissipates the heat into the surrounding outside air. This is especially likely with larger buildings or building complexes. The evaporative chiller is often called a *cooling tower.* It is that large box with louvers exhausting humid air or even mist that you so often find next to an office building. Some cooling towers may function for several buildings. There is a constant loss of water due to evaporation, so water must be added. Unfortunately, this means that dirt and minerals are also added, and left behind when the water evaporates. These must be drained off through a small valve in the base of the cooling tower called a *blowdown.*

On the evaporator side of the cycle, the coil takes the heat from water or air that is typically brought down to 50–55°F and then circulated around the building. This is the cooling loop for the building. In the seasons when it is cool enough outside, the outside air may be used directly and the refrigeration cycle

shut off. In some climates, cool water from a clean pond may be used in early winter and late spring, instead of chilled refrigerated water, or this pond water may be use for cooling the condenser without using an evaporative chiller. Such seasonal adjustments in the source are often referred to as an *economizer cycle.*

Heat Pump

Now, what would happen if we ran the entire system in reverse? That is to say, what if we refrigerated the outside, and collected the heat inside the building? Note that the refrigeration cycle does not create heat or make it disappear, but rather just moves it around. However, it does so very effectively. In expending 1 Btu of energy, we can actually move 2 to 4 Btu's. Normal boilers and furnaces run at about 80 percent efficiency. If we use a refrigeration cycle to heat a building, we can get the equivalent of 300 percent because we have *moved* heat energy from the outside to the inside of the building, in addition to the energy expended. We are not really creating energy, so we cannot call this the efficiency, but rather the *coefficient of performance (COP).* The definition of the two are basically identical. The difference is that the COP includes the heat delivered from the outside.

Efficiency = energy delivered / energy used.

COP = energy delivered / energy used.

Typical COP's vary between 2 and 3, with 2.3 being common. If the compressor is placed inside the building, much of the energy dissipated in friction, etc., is also gained as heat for the building.

There is some concern that all of the Freon (or any CFC) used in a refrigeration cycle will eventually leak into the atmosphere. It has been found that CFCs destroy ozone at very high altitudes. Enough ozone has already been destroyed so that there is a hole in the ozone layer over the South Pole that appears every spring. Several chemical companies are now working on alternative gases (typically hydro fluorocarbons or HFCs) for use in the refrigeration cycle. Over the long term, this is absolutely essential.

SYSTEM DISTRIBUTION TYPES

There is much more variation in distribution systems than in plant systems. However, there are only three basic categories: electrical, hydronic, and forced air.

Electrical Systems

Electrical systems are the simplest and the lowest in first cost, and generally the most expensive in life cycle costs. They are justified only in very mild climates where the system is off most of the time. There are two categories of electrical systems: radiant systems, which consist of radiant panels or wires embedded in the ceiling, and baseboard heaters, which heat up and cause convective air circulation in a room.

The advantages of the radiant systems are that the system only needs to be turned on in the rooms currently occupied, and that only objects, such as people, are heated, not the air (remember radiant heat transfer). However, since electricity is often generated from combustion at a low efficiency, this is a wasteful and often expensive way to use energy.

Hydronic Systems

Many hydronic systems are also radiant systems. Hot water or steam is circulated through registers or even through pipes embedded in the floor, which then radiate the heat into the space. Baseboard heaters using hot water or steam are also common. Hydronic systems may also be combined with forced air systems. For example, the hydronic system may bring hot and chilled water to each zone, and then an air circulating unit within the zone uses the water to heat or cool the air at that point, which is then blown into the room.

There are several different loop patterns to the piping in hydronic systems.

Single pipe is exactly what the name implies: a single supply and return pipe, which may be run in series or partly parallel. The hot water is circulated through each register (or fan coil unit, etc.) and back to the pipe. The first register is quite hot, and the temperature decreases with each successive register. This system has a low first cost, but may only be extended a very limited distance, since the water would be quite cold by the last register if there were more than five.

A *two pipe* or parallel system uses separate supply and return pipes. The water is not mixed back into the supply pipe after it has been through the register. Thus, the supply water for each register is at the same high temperature, since it hasn't been adulterated by cooler return water.

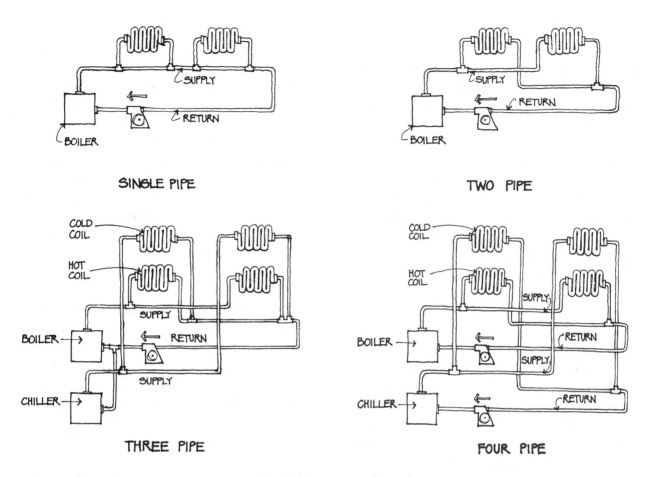

SINGLE PIPE

TWO PIPE

THREE PIPE

FOUR PIPE

HYDRONIC SYSTEMS

If both heating and cooling are desired, a separate two pipe system may be provided for each, resulting in a *four pipe system*. The temperatures available at each fan coil unit are fairly constant. Four pipe systems are used with two separate heating and cooling registers, rather than by mixing hot and cold water in a single register. The return pipes are kept separate.

A *three pipe* system mixes both the hot and cold water in a common return pipe. This saves on piping costs, but is more expensive to operate because all of the water must be heated or cooled from the middle temperature instead of having relatively hot and cold water in the separate return pipes.

Forced Air Systems

Forced air systems distribute heated or cooled air around the building using *supply ducts*. The air from the spaces may be returned through return ducts, or may be returned using the space between the suspended ceiling and the floor or roof above as one huge slow-moving air duct, called a *plenum*. In that case, the supply ducts may also be inside the interstitial space between ceiling and the floor above, but the supply air is sealed within the supply ducts and therefore does not mix with the return air. In some instances, the occupied spaces are used as the return plenum, through which the air slowly moves back toward the equipment room.

Doors have registers in them, and sometimes there is a *cold air register* between two floors, so named because return air to a residential furnace is cooler than the supply air. These registers allow the return air to move from space to space while maintaining visual privacy.

The air brought into the plant is usually a mixture of air returned from the building and fresh air brought in from outside the building. This means that there must be a *fresh air intake*, preferably away from the cooling tower and other exhausts, and a duct from the intake to the mechanical equipment room. The outside air must be warmed or cooled, and the return air must be cleaned or filtered. The number of air changes in a room is specified by code or by good practice, and must be provided even if that volume of air is not necessary for meeting the thermal load.

The air is pushed from the plant out through the ductwork by the supply fan, which must have sufficient flow volume to provide enough air and sufficient pressure to overcome the friction in the duct. This pressure is usually expressed in terms of static head of water, which is the height of a column of water that could be lifted by the pressure. (See also Lesson Two.)

If we want to eliminate infiltration in a building, we positively pressurize the building, by running the supply fan at a rate greater than the sum of the return air fan rate and the leakage rate of the building.

The temperature of the air as it leaves the equipment room is called the equipment temperature or the *deck temperature*. For example, the *cold deck* might be 55°F. As the air moves through the supply ducts, it tends to warm up somewhat before arriving at the room. Insulating the supply ducts keeps this to a minimum. In a supply duct, exterior insulation is used for temperature control, while interior insulation is usually for sound control.

Fans must be isolated from the floor on which they sit and from the ducts to which they are connected so that their vibrations are not transmitted as noise throughout the building. This is usually done by mounting each fan on springs, which are mounted on rubber pads, with the entire assembly placed on a massive concrete pad. The duct connections are rubber or even fabric, so that the pressure and flow are maintained, but the connection is flexible.

The simplest forced air system is the *single duct* system, often found in residential applications. The amount of air sent out through the ducts is constant, which is called *constant volume*. The furnace simply runs until the temperature reaches the thermostat setting. It is impossible to heat one room and cool another simultaneously. All rooms may be heated, or all rooms may be cooled if there is a central air conditioning unit. The rate of heating may be varied from room to room by putting a *damper* on each *diffuser*, which is simply a metal flap that shuts off or reduces the airflow coming through the louvers or grill at the room. If the air needed for ventilation is greater than the air needed for heating, there is a problem.

A variation of this system is the electric reheat system, in which all of the air is cooled, and an electric resistance heater is placed in the duct upstream of each diffuser. In rooms that need to be heated, the cooled air is simply reheated by the coil. Needless to say, this is not terribly efficient, and is appropriate only for climates where heating is rarely necessary.

The *double duct*, or *dual duct*, system is a combination of two single duct systems, one carrying heated air and one carrying cooled air. This requires twice as much duct space, but

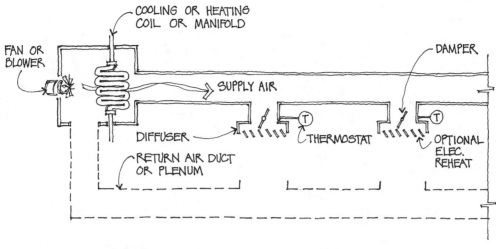

SINGLE DUCT CONSTANT VOLUME SYSTEM

can heat one room and simultaneously cool another. The airflow rate in the ducts is constant, but the amount pulled from each duct at each room is controlled by dampers, and mixed in a *mixing box* at a ratio controlled by the thermostat to provide the appropriate temperature. This system requires more space than any other system, but it is ideal for linear buildings with many different thermal conditions. The two ducts

are usually run parallel above a central corridor, with each zone connected to both ducts.

In order to reduce the amount of space taken up by ducting and the cost of the duct work, the *multizone* system was developed. It is similar to the double duct system, only the mixing boxes are inside the mechanical equipment room, and pre-mixed air is sent out to each zone. If the

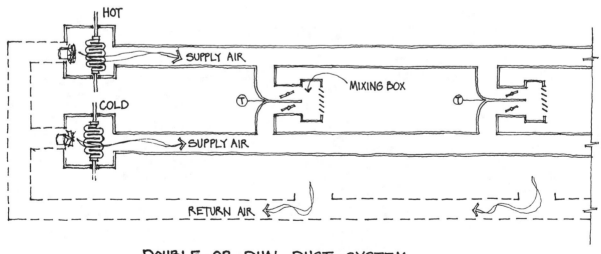

DOUBLE OR DUAL DUCT SYSTEM

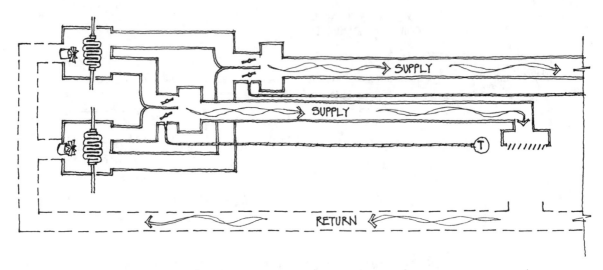

MULTIZONE SYSTEM

building is square in plan, and there are few zones, this system is efficient. However, if the building has many zones, then many small ducts must be sent out rather than two large ducts, and the system becomes much less economical.

The *fan coil* system is one of the most efficient systems that can heat and cool simultaneously. A constant volume of cleaned and conditioned air is supplied from the plant in a single duct.

A chilled water and a hot water pipe are also supplied. At each zone there is a unit with a fan and two coils. The air is blown over the coils into the room. Hot water may be run in the hot coil if the room needs to be heated, cold water in the cold coil if the room needs to be cooled, and no water at all if the room just needs air for ventilation. Unfortunately, the first cost is rather high, since there is a great deal of plumbing (it is essentially a three or four pipe system) and as much sheet metal as for a single duct system.

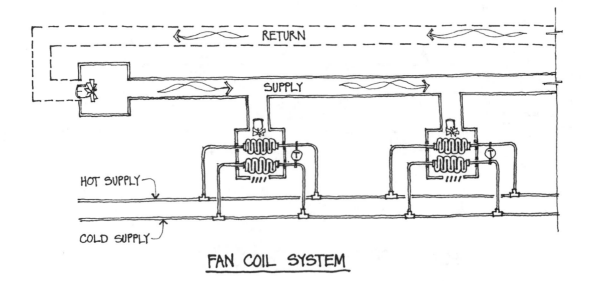

FAN COIL SYSTEM

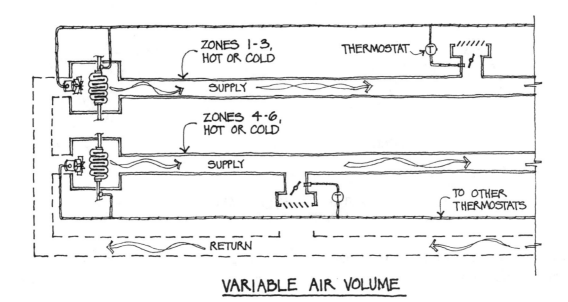

VARIABLE AIR VOLUME

The most common efficient system is the *variable air volume (VAV)* system. This is a single duct system, or there may be three or four separate single duct systems, each of which serves a zone, which may be composed of several rooms. Instead of a constant flow volume, the flow rate may be varied. It is impossible to simultaneously heat one room and cool another within the same zone, but different zones can be handled differently. Thus, one zone may be heated while another is being cooled.

All of the air going to a given zone is at the same temperature, and the *amount* of heating or cooling delivered is determined by the *volume* of air delivered. The temperature of the air in the duct or the overall air flow rate may be constantly adjusted so that the coolest, or hottest, room is just barely taken care of, and thus the system runs at high efficiencies.

The term *unitary systems* covers many kinds of systems. If the air comes directly from the outside through the unit and into the room, then it is a form of unitary system. There is one unit for each zone, often placed on the roof above the zone, or in a permanent cabinet along a wall. The unit is usually self contained, requiring only electricity, but it may be connected to a chilled or heated water supply which originates elsewhere. Unitary systems are employed most often where the building is spread out and ducting would be impractical or costly because of the length of the run and the temperature drop, and/or lack of space. Unitary systems are also one of the systems used when each zone must have a separate utility bill.

A *heat pump system* is a term applied to a group of heat pumps that serve a building. The system can be quite efficient, although the first cost may be higher. Water is circulated throughout the building and is called the *heat sink*. Each zone has its own heat pump and a fan and short ducts to recirculate air within that zone. The heat pump either removes heat from the water and adds it to air blown into the zone, or removes heat from the zone air, adding it to the water. At certain times of the year, this all balances out, and the water just needs to be circulated. If all the heat pumps are being used for cooling their rooms, the water temperature begins to rise and a chiller in the mechanical equipment room cools it back down. If the water temperature

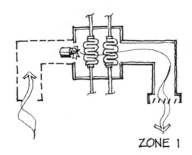

ZONE 1

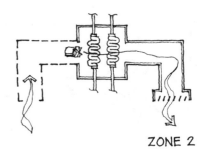

ZONE 2

UNITARY SYSTEMS

begins to drop, it is routed through a boiler, which heats it back up.

There are several other terms that are used in forced air systems. *Induction* is the term applied to any system in which a small amount of supply air at a very high velocity is delivered to a box-like unit and mixed with air brought in from the room, which induces a greater airflow than just the supply air delivered to the room.

There are various types of fans. We are most familiar with the bladed fan, similar to a propeller in form and function. When moving

large amounts of air, a *centrifugal fan*, sometimes called a *squirrel cage blower*, is often used.

In addition to heating, cooling, or dehumidifying the air, the air must be cleaned prior to circulation or recirculation. *Fibrous filters*, similar to home furnace filters, remove a great deal of the dust and lint, and must be replaced regularly. *Electrostatic filters* are more expensive, but produce less resistance to air movement. They are composed of two sets of charged plates that attract dust particles, which are then cleaned or washed off. *Activated charcoal filters* remove odors and many chemicals from the air, but are

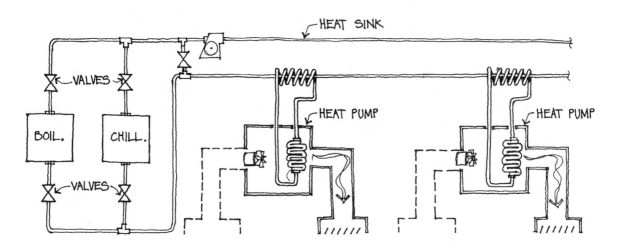

HEAT PUMP SYSTEM

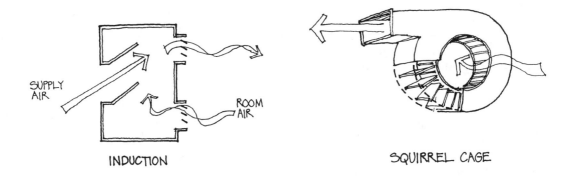

INDUCTION

SQUIRREL CAGE

INDUCTION UNIT AND SQUIRREL CAGE BLOWER

used only when this is necessary. Such filters present a significant resistance to airflow, require very low velocities, and must also be replaced regularly.

PLANT AND DUCT SIZING

Architects and architectural candidates generally do not need to know how to design the mechanical equipment in detail. However, he or she should know how much space it will require. There are two parts to that question: How much floor space will the mechanical equipment in the plant take? And, how much space will the distribution system take in the building cross section?

Plant Sizing

Although it varies with the complexity of the system, the floor space required for the mechanical equipment is usually 5 to 10 percent of the total floor area of the building. This space may be provided in the basement or in a penthouse on the roof. In either case, the architect must provide proper access to the space, since the equipment will eventually wear out and require maintenance or replacement. In a high-rise office tower, one out of every 15 or 20 floors

may be given over entirely to mechanical equipment.

The capacity of the plant must be sufficient to take care of all of the loads experienced on the design day, based on the efficiency of the distribution system. Heating loads may be expressed in thousands of Btu's per hour or kBtuh. Cooling loads are often expressed in tonnage. A *ton* of cooling is equivalent to *12,000 Btuh*, which is the rate of heat transfer that would melt a ton of ice over a 24-hour period.

System Sizing

As far as the distribution system is concerned, no simple rules of thumb apply. Different systems require different amounts of space, and the layout of the system determines where that space is needed. The forced air systems are the most space consuming. In forced air systems there is a relationship among space, volume, velocity, and noise. Higher flow volumes require either greater cross-sectional area for the duct or higher velocities. Higher velocities produce greater friction, which often creates a noise problem.

The numerical relationship is expressed by the formula:

$$A = 144\ Q_{cfm}/v$$

where

Q_{cfm} = the flow rate in cubic feet per minute (remember that infiltration used Q measured in cubic feet per hour)

v = the velocity measured in feet per minute

A = the cross sectional area of the duct in square inches

Q_{cfm} may be determined from the formula:

$$Q_{cfm} = q_{tot}/1.08\,(T_{eq} - T_i)$$

where

q_{tot} = the total thermal load in Btuh

T_{eq} = the temperature of the air supplied in the duct (55°F for cooling or 140°F for heating)

T_i = the desired interior temperature for the room

Duct Sizing

We can size the ducts by choosing an appropriate velocity and checking whether the resultant duct size fits into the ceiling cavity, or by choosing the duct size and checking if the resultant velocity will cause excessive noise or too much friction for the fans to overcome.

Appropriate velocities range from 300 fpm for quiet and barely noticeable velocities at the diffuser to 2,000 fpm in a duct. Some large office buildings use vertical ducts or ventilation shafts with velocities in the 10,000 fpm range.

Duct sizes are given in square inches of cross sectional area. For example, a 12×12 inch duct is 144 in², and a 20×7 inch duct is 140 in². The most efficient duct for a given area is the one with the least perimeter, since that results in the least resistance to air movement and the least friction. For this reason, circular cross sections are the optimum shape, if there is sufficient space. Ducts are often specified in terms of their

equivalent circular diameter (the circular diameter that would result in the necessary cross sectional area).

Fan Sizing

The above equations can yield the duct size given the necessary flow rate and a desired velocity. There is a another factor that must be considered: the friction of the air traveling through the ducts. For this reason, the duct sizes are often determined from a graph that plots velocity, flow rate, and duct size, in addition to friction loss.

The friction loss is expressed in inches of water per 100 feet, also know as the *static head*. One inch of static head is the pressure required to support a one inch column of water. (Normal atmospheric pressure would support a 33 foot column of water.) This information is used in two ways. First, the rating of the fans in the mechanical room is compared with the friction loss through the system to make sure that they can provide sufficient pressure to overcome that loss and push the air to the farthest diffuser at the flow rate required. Secondly, if the friction loss becomes excessive, larger duct sizes can be chosen, or special fans specified.

ENERGY CODES

Energy codes of one sort or another are becoming increasingly common across the country. There are two basic types: prescriptive codes and performance codes. A *prescriptive code* specifies how to build a building, while a *performance code* states what the final result needs to be and how it will be measured, but does not specify how that result is to be achieved. There are various guidelines, as well. The best known is the *ASHRAE 90-xx* series, which covers suggested practice in the areas of external envelope, HVAC

TABLE 5.1 FRICTION LOSS

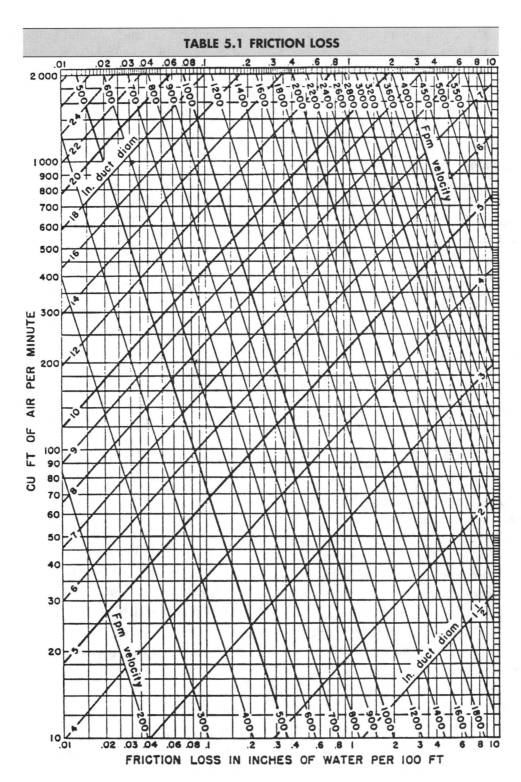

FRICTION LOSS IN INCHES OF WATER PER 100 FT

CU FT OF AIR PER MINUTE

Reprinted by permission from ASHRAE Handbook of Fundamentals.

equipment, water heating equipment, and electrical distribution. This is an example of the prescriptive approach. The *Building Energy Performance Standards (BEPS),* which have, at various times, been proposed for any work done or financed by the Federal government, specify an energy budget per square foot for various building functions. The budget varies according to climate (i.e., California is divided into 16 different climate zones) and the function of the spaces within the building. This is an example of a performance approach.

Most states use a combination of the two approaches. There is a prescriptive version, typically based on an *overall thermal transmission value (OTTV),* which is a weighted average U-value for all of the exterior surfaces of the building. This does not take any solar design or orientation factors into account, but is based on the "thermos bottle" concept: if it doesn't leak heat too badly, it's not a bad building.

Most states also allow a performance version of the same code. This is done by modeling an OTTV version of your building (same square footage, etc.) on a certified computer program, and then modeling the building as you actually designed it, using the same computer program. If your version performs as well or better than the OTTV version, your building is acceptable.

There are other versions, which trade points for meeting or exceeding the prescription in one area for points lost by not meeting the prescription in some other area.

BENCHMARKING

The U.S. Department of Energy provides "benchmark" information of total energy consumption in BTU's/SF for various kinds of buildings in the United States. These standards,

or benchmarks, can be useful in the measuring of energy efficiency standards for various types of buildings:

For example:

■ Average for all office buildings (pre-1990)	104.2
■ Average for all office buildings (1990–1992)	87.4
■ Average for all educational buildings (pre-1900)	87.2
■ Average for all educational buildings (1990–1992)	57.1
■ Average for all laboratory buildings (pre-1990)	319.2
■ Average for all health care buildings (pre-1990)	218.5

Source: U.S. DOE, Commercial Building Energy Consumption and Expenditures.

Benchmarking is a good way to alert the design team to the base energy standards for their design. It's a good place to start and ultimately a standard to beat. And, one can see from some of the comparisons (Office and Educational Buildings), that some energy efficiency is occurring, but we still have improvements that are needed with all buildings, especially in the area of Laboratory and Health.

COMMISSIONING

Commissioning is an organized process to ensure that all building systems perform interactively according to the intent of the architectural and engineering design, and the owner's operating needs.

Commissioning usually includes all HVAC and MEP systems, controls, ductworks and pipe insulation, renewable and alternate technologies, life safety systems, lighting controls and daylighting systems, and any thermal storage systems. Commissioning also verifies the proper operation of architectural elements such as the building envelope, vapor and infiltration control, and gaskets and sealant used to control water infiltration.

Commissioning is a process required for LEED certification, but is a recommended procedure for any building involved with sustainable design procedures.

Source: *Commissioning Requirements for LEED Green Building Rating, Version 8.* February 5. 1999.

Page 71, *The HOK Guidebook to Sustainable Design,* Sandra Mendler & William Odell, John Wiley & Sons, Inc., New York, 2000.

INNOVATIVE TECHNOLOGIES

Besides the aforementioned issues of solar design, improved lighting systems, improved HVAC systems, and improved building massing and envelope design, there are several "innovative technologies" that the architect can offer to their project team for consideration.

1. *Ground Water Aquifer Cooling and Heating (AETS)*

   One alternative to full air-conditioning with chillers, which make heavy demands on electricity, is the aquifer thermal energy storage that uses the differential thermal energy in water from an underground well to cool a building during summer and heat a building in the winter.

This is an efficient and relatively low cost system, but it may require approval from the local environmental authority before installation.

2. *Geothermal Energy*

   Where appropriate, heat contained within the earth's surface causes macro-geological events (such as underground geothermal springs or lava formations) that may be tapped to produce heat for adjacent structures.

   In select locations this heat energy can be transferred and conveyed to supplement a building's heating demand.

3. *Wind Turbines*

   Small-scale wind machines used to generate electricity can be mounted on buildings or in open space nearby. These systems have the advantage of being relatively cost effective; a tested and established technology; systematically started-up; and have a relatively high output.

   These systems have the disadvantage of needing a relatively high mast; requiring substantial structural support; potentially causing noise problems; and being visually intrusive.

4. *Photovoltaic (PV) Systems*

   The basis of the PV systems is the concept that electricity is produced from solar energy when photons or particles of light are absorbed by semi-conductors.

   Most PV systems are mounted to the building (either on the roof or as shading devices above fenestration). Currently, PV systems are not cost effective. But with promised government subsidy necessary to achieve an economy of scale, PVs may be a viable method of electrical production in the United States, Japan, and Germany in the near future.

5. *Fuel Cells*

Even though Sir William Grove invented the technology for the fuel cell in 1839, it has only recently been recognized as a potential power source for the future. The fuel cell claims to be the bridge between hydrocarbon economy and the hydrogen-based society.

Fuel cells are electrochemical devices that generate direct current (DC) electricity similar to batteries. But, unlike batteries, they require a continual input of hydrogen-rich fuel. In essence, the fuel cell is a reactor that combines hydrogen and oxygen to produce electricity, heat, and water. It is clean, quiet, and emits no pollution when fed directly with hydrogen.

At the moment, the fuel cell technology is still not cost effective for the commercial building market. But, there seems to be a general feeling that hydrogen-based energy reactors will soon be an optional energy source.

6. *Biogas*

Biogas is produced through a process that converts biomass, such as rapid-rotation crops and selected farm and animal waste, to a gas that can fuel a gas turbine. This conversion process occurs through anaerobic digestion—the conversion of biomass to gas by organisms (like bacterium) in an oxygen free environment.

Biogas has several advantages: it is a relatively high energy production; it lends itself to both heat and power production; it creates almost zero carbon dioxide emissions; it virtually eliminates noxious odors and methane emissions; and it protects ground water and reduces the landfill burden.

7. *Small Scale Hydro*

Harnessing the energy from moving water is one of the oldest energy production systems in the world. In some locations, small scale hydro power is a efficient and clean source of energy and is devoid of environmental penalties associated with large scale hydro projects. It allows small scale, local energy production, with relatively low cost.

8. *Ice Storage Cooling Systems*

One of the problems for energy supply companies is that the highest demand for electricity often coincides with the highest cooling demand.

The utilities would prefer to "flatten the curve" (to even out or flatten the measure of average daily energy demand). The less number of peaks (high points of energy demand), the less the utilities have to supplement their power supply with expensive, supplemental fuels.

One way to reduce this peaking problem is to supplement a building's cooling capacity with an ice storage system.

An ice storage system has three components: a tank with liquid storage balls, a heat exchanger, and a compressor for cooling.

The essence of the ice storage system is that the chilling and freezing of the ice balls occurs at night (when the cost of energy is lower due to lower demand). During the day, the cool temperatures, stored in the ice, are transmitted into the building's cooling system.

Source: *Sustainability at the Cutting Edge*, Peter F. Smith, Architectural Press, An Imprint of Elsevier Science. Linacre House, Jordan Hill, Oxford OX2 80P, 2003.

SUMMARY

We have considered the different approaches to heating and cooling buildings, including various types of plants and distribution systems. We discussed the refrigeration cycle and different ways that it may be used. We looked at different distribution systems: electric, hydronic, and forced air, and what is involved in sizing forced air systems.

The terms of HVAC equipment are no longer foreign to us, and we have some idea of how energy codes are developed.

This information, along with the information on thermal processes, human comfort, climate, and solar design covered in the previous lessons represent the basic concepts of HVAC on which you may be tested in the examination, and even more important, what we really ought to know to design good buildings.

LESSON 5 QUIZ

1. Which of the following systems can be combined with a four-pipe hydronic system?

    A. Double duct system
    B. Multizone system
    C. Fan coil system
    D. Unitary system

2. All of the following are capable of reversing the otherwise invariant flow of heat from hotter to cooler objects, EXCEPT

    A. the heat pump system.
    B. the variable air volume system.
    C. the refrigeration cycle.
    D. the evaporative chiller.

3. Which of the following systems can provide simultaneous heating and cooling to different rooms within the same zone?

    I.   Single duct constant volume with electric reheat system
    II. Double duct system
    III. Fan coil system

    A. I and II   C. I and III
    B. II and III   D. I, II, and III

4. Which of the following uses the latent heat of evaporation to transfer heat?

    A. Evaporative chiller
    B. Multizone system
    C. Unitary system
    D. Variable air volume system

5. Which of the following systems can be used if each zone must have a separate utility bill?

    I.   Unitary system
    II. Heat pump system
    III. Double duct system

    A. I and II   C. I and III
    B. II and III   D. I, II, and III

6. Which system uses a heat sink and often balances itself without utilizing a boiler or chiller?

    A. Unitary system
    B. Multizone system
    C. Heat pump system
    D. Variable air volume system

7. Which part of the refrigeration cycle is the hottest?

    A. Evaporator   C. Condenser
    B. Compressor   D. Valve

8. Which of the following hydronic systems are parallel?

    I.   One pipe
    II. Two pipe
    III. Three pipe
    IV. Four pipe

    A. I and II   C. III and IV
    B. II, III, and IV   D. I, II, III, and IV

9. Which of the following hydronic systems can provide both heating and cooling simultaneously?

    I.   One pipe

    II. Two pipe

    III. Three pipe

    IV. Four pipe

    A. I and II

    B. II, III, and IV

    C. III and IV

    D. I, II, III, and IV

10. Where does return air go?

    A. To the plant

    B. To the outside of the building

    C. To the room that requires air at that temperature

    D. To the exhaust flue

ELECTRICAL SYSTEMS

INTRODUCTION

Although the use of electrical power has little effect on the form or shape of buildings, it is of tremendous importance when it comes to the tasks performed within buildings. In fact, without the electricity that powers its typewriters, air conditioning, lighting, and computers, the modern office building would be unable to function. The architect should therefore have an understanding of the basic principles of electrical systems, and the Architect Registration Examination tests this understanding.

BASIC PHYSICS

Because we cannot see electricity, perhaps the best way to understand it is by looking at the flow of water as an analogy.

The three basic factors in electricity are *potential, current*, and *resistance*. The chart on the following page shows the analogous factors in water flow.

	Water	**Electricity**
Potential	Height, pressure or pressure difference (feet or psi)	Voltage V (volts)
Current	Flow (gallons per minute)	Current I (amperes)
Resistance	Resistance to flow (inches per 100 ft.)	Resistance R (ohms or Ω)

The formula which relates these factors to each other is called *Ohm's Law*:

$$I = V/R$$

where

   I = current, measured in amps
   V = voltage, measured in volts
   R = resistance, measured in ohms

The formula makes sense, for both the analogous water flow and electricity. The greater the voltage, the greater the current (the greater the water pressure, the greater the water flow). Conversely, the greater the resistance, the smaller the current.

As with water, there may be several resistances in the flow path, or there may be parallel paths with different resistances and flow rates in each path. In electricity, these two conditions are called respectively *series resistances* and *parallel resistances*. The effective resistance of each may be calculated.

With series resistances, the effective total resistance is the sum of all of the resistances:

$$R_{tot} = R_1 + R_2 + R_3 + \ldots + R_n$$

With parallel resistances, the effective total resistance is expressed by the formula:

$$1/R_{tot} = 1/R_1 + 1/R_2 + 1/R_3 + \ldots + 1/R_n$$

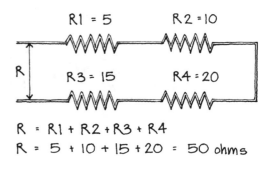

$$R = R1 + R2 + R3 + R4$$
$$R = 5 + 10 + 15 + 20 = 50 \text{ ohms}$$

SERIES RESISTANCES

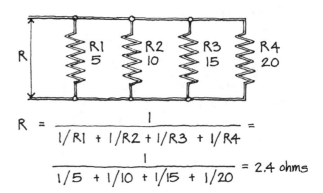

$$R = \cfrac{1}{1/R1 + 1/R2 + 1/R3 + 1/R4} = \cfrac{1}{1/5 + 1/10 + 1/15 + 1/20} = 2.4 \text{ ohms}$$

PARALLEL RESISTANCES

TRANSMISSION AND USAGE

There are several schemes for using and transmitting electricity, the most common of which are described below.

Direct Current

Direct current (DC) means current that flows only in one direction, with constant voltage. This is typical for low voltage applications, as with batteries, for example. Low voltages are less dangerous, in general, than higher voltages, because they result in less current through a given resistance. The general equation for power in a DC circuit is:

$$P = V \times I$$

where

  P = power in watts
  V = voltage
  I = the current in amps

For example, for a 12 volt battery connected to a 4 ohm resistor, the current is

$$I = V/R = 12v/4\Omega = 3 \text{ amps}$$

$$P = V \times I = 12v \times 3 \text{ amps} = 36 \text{ watts}$$

Alternating Current

Alternating current (AC) is based on the concept that electricity has nearly no inertia, and therefore the direction of the flow can be reversed very rapidly by reversing the voltage. If we plot the voltage in such a circuit, it results in a sine wave.

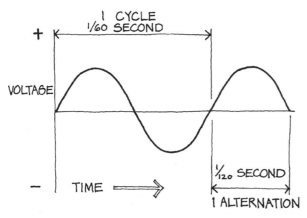

ALTERNATING CURRENT SINE WAVE

The current flow may lag behind the voltage reversal, and therefore the amount of power is not as simple to calculate as with a DC circuit. The *power factor* is the cosine of the angle between the voltage wave and the resultant current wave. It ranges from 0.0 to 1.0, although it is usually expressed as a percentage, i.e., from 0 to 100 percent. In a single-phase circuit, the power formula is:

$$P = V \times I \times PF$$

where

  P = power in watts
  V = voltage in volts
  I = current in amps
  PF = power factor in decimal form

There is also a three-phase version of alternating current, in which there are three different circuits, each 120° out of phase with the others, and one neutral or ground circuit. The formula for power in a three-phase circuit is:

$$P = V \times I \times PF \times \sqrt{3}$$

where

  P = power in watts
  V = voltage in volts
  I = current in amps
  PF = power factor in decimal form

For example, if a three-phase motor draws a current of 7 amps at 240 volts and the power factor is 0.8, the power

$$\begin{aligned} P &= V \times I \times PF \times \sqrt{3} \\ &= 240 \text{ v} \times 7 \text{ amps} \times 0.8 \times 1.73 \\ &= 2{,}325 \text{ watts} \end{aligned}$$

When we deal with a large amount of power, we commonly use the unit *kilowatts* (1,000 watts) or even megawatts (1,000,000 watts). Thus 2,325 watts is the same as 2.325 kilowatts.

ELECTRICAL EQUIPMENT

A motor is a machine that converts electrical energy into mechanical energy. The converse is a generator, a machine which converts mechanical energy into electrical energy.

Rotating a wire loop between two magnetic poles will generate a current. This is the basic principle behind *generating* electricity.

Conversely, running a current through a wire wrapped around an iron core creates a magnetic field. Since magnets may attract or repel each other, the magnetic field can create motion. This is the basic principle behind electric *motors* and electric *solenoids*.

A solenoid is a wire wound spirally around an iron core to produce a magnetic field and that is used as an electromagnetic switch.

Generation of Power

A single-phase *alternator* is the most basic form of power generation. The resultant power is AC current, in which the time interval from peak to peak of the voltage sine wave is based on the number of revolutions per minute (rpm) of the shaft on which the wire loop is mounted. This is usually 60 rpm, which results in a peak to peak time (one cycle) of 1/60 of a second, or 60 cycles per second or 60 *Hertz*, the typical AC power frequency in the United States. In Europe, 50 Hertz is the common frequency for power. In the United States, the common household voltage is 110 volts, while in Europe it is 220 volts. This is the magnitude from the bottom to the peak of the voltage sine wave.

Three-phase power is generated by putting three loops on the shaft and keeping them as separate circuits. If the loops are evenly spaced around the circumference of the shaft, the sine wave current generated is shifted by 1/3 of a cycle (or 120 degrees) between each circuit.

The resulting currents can be represented by three separate sine waves. Note that if only one of the circuits is connected, normal single-phase current is delivered.

Transformers

Devices that change the voltage of an AC circuit to a higher or lower value are called transformers.

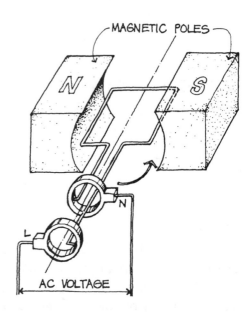

SINGLE PHASE ALTERNATOR

Basically, a transformer consists of an iron core on which two separate coils of wire are wound. The coil (also called a winding) with the greater

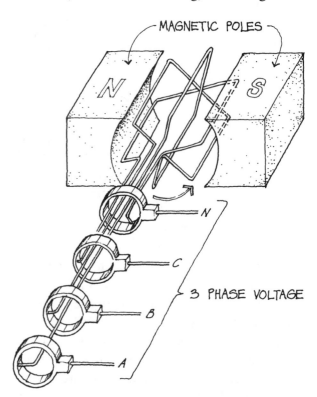

THREE PHASE ALTERNATOR

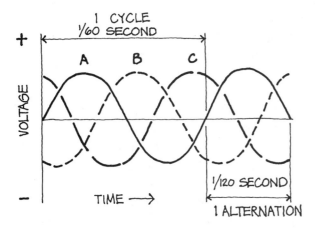

THREE PHASE CURRENT

number of turns will have a higher voltage, and the one with fewer turns will have a lower voltage. We can run current through the winding with the greater number of turns and produce a lower voltage in the other winding, or vice versa. While a transformer changes the voltage in a circuit, it has practically no effect on the total power in the circuit. Transformers are used to step up voltage in order to transmit power over long distances without excessive losses, and subsequently step down voltage to more usable household levels.

Transformers are called *step up transformers* when they increase the voltage and *step down transformers* when they decrease the voltage, which is usually the case in buildings.

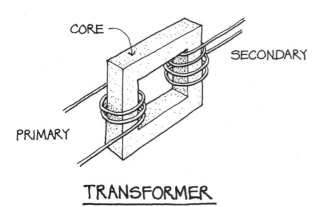

TRANSFORMER

Transformers waste surprisingly little energy, but what is wasted turns into heat, which must be dissipated by the transformer. The thermal rating of the transformer is the product of the Voltage and the Amperage or VA. This is not exactly the same as power ($V \times I \times PF$) because it represents how much heat the transformer can handle without melting or exploding, not power being delivered. Again, in the scale of buildings, one thousand VA is a handy unit, and is abbreviated KVA.

In small transformers, the wires are simply insulated by rubber or vinyl or other usual insulating materials. In larger transformers, however, the wires are insulated by a fluid that is resistant to the flow of electricity and also capable of withstanding the high internal temperatures, rapidly conducting the heat away from the windings.

Transformers should be properly ventilated, since their thermal rating is based on the assumption that the surrounding air will be able to remove heat at a certain rate from the cooling fins on the transformer. If a large transformer overheats, it may explode. The fluid that insulates the electricity and cools the transformer is often toxic, and the explosion may vaporize it or spray it over adjacent surroundings or people.

For these reasons, transformers must either be placed outside the building, or within the building inside a fireproof vault. In addition, transformers make a certain amount of noise (hum) and it is desirable to isolate that as well.

Transformer Connections

The *primary* winding is the winding that is used for input in a transformer, while the *secondary* winding is used for output. The secondary may be divided into segments, so that the output voltage depends on which segments are used. Single-

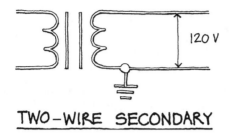

TWO-WIRE SECONDARY

phase transformers may have two- or three-wire secondaries.

A *two-wire secondary* has one wire grounded, which then becomes the neutral.

A *three-wire secondary* consists of two segments. One lead is at one end of the secondary. A second lead is connected to the midpoint of the secondary and grounded, and the third lead is connected to the other end of the secondary. If 240 volt output is desired, the first and third lead are used, and the midpoint lead is ignored. For 120 volts, the first and midpoint leads or the midpoint and third leads are used.

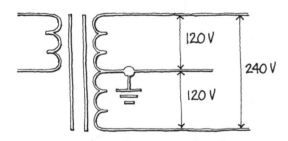

THREE-WIRE SECONDARY

Similarly, three-phase transformers may have multiple leads on the secondary winding, and even different configurations to both the primary and the secondary.

The two basic types of connections are called wye (shaped like the letter Y) or delta (shaped like the Greek capital letter delta Δ, or a triangle). The wye is sometimes referred to as a "star" because the neutral contact (at the crotch of the Y) forms the center of a three pointed star.

The illustration on the following page shows five basic methods for connecting three-phase transformers (the iron core symbols have been omitted from the illustration).

When a neutral connection is desired, it may be taken from the center point of a wye, or from the midpoint of one of the secondary windings of a delta. The illustration on page 120 shows alternative representation of delta-delta and delta-wye transformers with neutral connections added. The parallel lines represent the iron cores, and the neutral wire is shown grounded for safety, which is typical.

Primaries are usually connected delta, and seldom have a ground connection. Since the wye connection is symmetrical, the voltage from each of the three-phase wires (A, B, and C) to the neutral is the same, and is equal to the voltage from line-to-line divided by $\sqrt{3}$ (or 1.73). Typical system voltages are 120/208 and 277/480, where the lesser voltage is the line-to-neutral voltage and the greater voltage is the line-to-line voltage. Note that in both cases the ratio of the voltages is 1.73.

Where a neutral is applied to a delta, the voltage from two of the phase wires (B and C) to the neutral is one half of the phase-to-phase voltage. The voltage from the third phase (A) to the neutral is 0.866 times the phase-to-phase voltage, but no loads are ever connected between these two points.

Residences are limited to 120/240 volt single-phase systems, which come from a three-wire secondary, and larger loads, such as the electric range, air conditioner, refrigerator, and other semi-permanent connections, use the line-to-line voltage of 240 volts. The receptacles and light switches use the line-to-neutral voltage of 120 volts.

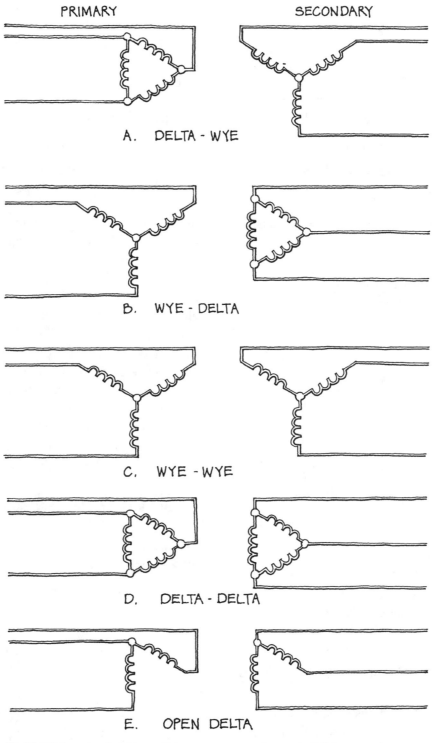

PRIMARY   SECONDARY

A.   DELTA - WYE

B.   WYE - DELTA

C.   WYE - WYE

D.   DELTA - DELTA

E.   OPEN DELTA

THREE — PHASE TRANSFORMER CONNECTIONS

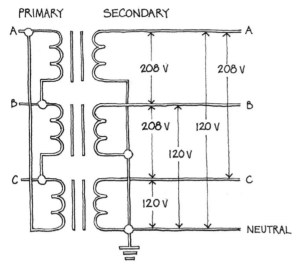

DELTA - WYE SYSTEM
120/208 V 3 PHASE 4 WIRES
277/480 V HAVE SAME CONNECTIONS

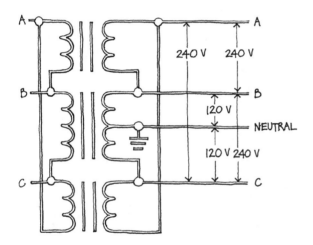

DELTA - DELTA SYSTEM
120 V/240 V 3 PHASE 4 WIRES
WITHOUT THE NEUTRAL, THIS CAN BE
240 V 3 PH 3W OR 480 V 3 PH 3W

3·PHASE TRANSFORMER SCHEMATICS

Small commercial buildings usually receive 120/240 volt single-phase or 120/208 three-phase service, but larger commercial and industrial applications require the higher voltage three-phase services, such as 277/480 volts or 2,400/4,160 volts.

To summarize, the voltages in Table 6.1 are in common use as utilization and distribution voltages. Even higher voltages are used for distribution and transmission, but will not be discussed here.

Electric Heaters

Electric heaters, the heating coils in furnaces, and even the coils in hair dryers that warm the airflow are all basically the same thing: a length of stainless steel wire formed into a coil and supported on insulated prongs. The wire is simply a resistance to the current, which generates heat. In a motor, this would be wasted energy, but in a heater, it is exactly what we desire. Electric heaters are 100 percent efficient, in that every bit of electrical energy is converted to heat. However, since much of the electricity in this country is generated by using heat to generate steam, which generates electricity at about a 30 percent efficiency, using electricity to heat a space is very wasteful in the broad ecological sense. On the other hand, using electric heat as a *radiant heat* source (see Lessons Three through Five) is efficient because it heats only people, and not air.

TABLE 6.1—TYPICAL SECONDARY VOLTAGES			
Number of Phases	Voltage	Connection Name	Number of Wires
1	120	2 wire	2
	120/240	3 wire	3
3	240	Delta	3
	120/240	Delta	4
	480	Delta	3
3	208	Wye	3
	120/208	Wye	4
	480	Wye	3
	277/480	Wye	4
3	2,400/4,160	Wye	4
	4,160	Delta	3
	4,160	Wye	3

Electric Lighting

We have discussed lighting at some length. Lights are often grouped into circuits that may be switched on and off from a central panel in commercial installations, or from wall switches next to doors in residential applications. When there are two or more doors into a room, there is often a light switch placed at each door. Such switches are called *three-way* switches, in which either switch controls the lighting circuit. Thus, the on position of a switch may be either up or down, depending on the position of the other switch. When more than two switches are necessary, two of the switches must be three-way switches and the remaining additional switches must be *four-way* switches.

Motors

Only four types of motors are in general use. The *DC motor* is used for small scale applications and for elevators, where continuous and smooth acceleration to a high speed is important. *Single-phase AC motors* are used in many sizes and shapes, typically 3/4 horsepower or less. Larger motors are typically *three-phase induction motors.* Such motors remain constant in rpm, unless heavily overloaded. Typical power factors are in the range of 0.7 to 0.9. They are characterized by extreme reliability. The only other motor commonly available is the universal motor, which runs on either DC or AC current, but which varies in speed based on the load. These are often found in mixers, hand drills and similar appliances.

All motors should be protected against overload by *thermal relays* that shut off the power when any part of the motor or housing gets too hot.

Capacitors

The simplest capacitor is a set of two plates separated by a small insulating layer. Current is "stored" on one plate and at some point, the entire stored amount is discharged. Capacitors are used to improve the power factor in a circuit. This improves efficiency and overall performance.

Receptacles

A *receptacle* is commonly known as an *outlet,* or erroneously known as a wall plug. (A plug is actually what goes into the outlet.) In residential construction, outlets should be no farther than 12 feet apart. All outlets should be three-prong outlets, where the third prong is grounded. All of the outlets in a large room should not be on the same circuit. Thus, when a fuse or circuit breaker trips because of an overload, the room will not be plunged completely into darkness.

Panelboards

A panelboard is a set of fuses or circuit breakers that control the circuit loading in a building. It provides a central distributing point for branch circuits for a building, a floor, or a part of a floor. Each breaker serves a single circuit, and the overload protection is based on the size and current-carrying capacity of the wiring in that circuit. A building may have several panelboards and a main panel, with a disconnect switch for the entire building.

Wiring

The sizes of electrical wiring are standardized using *American Wire Gage (AWG).* No sizes smaller than *14 gage* should be used for building wiring.

Aluminum wiring has been discontinued for small gages (#4 or less) in most areas because of some questions about its oxidation, connection deterioration, and metal fatigue over a long period of time. Several major fires have been blamed on the use of aluminum, and copper wire is again the standard for branch circuits.

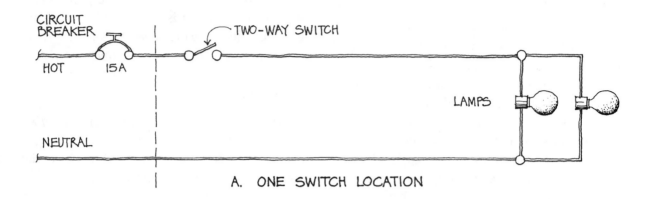

A. ONE SWITCH LOCATION

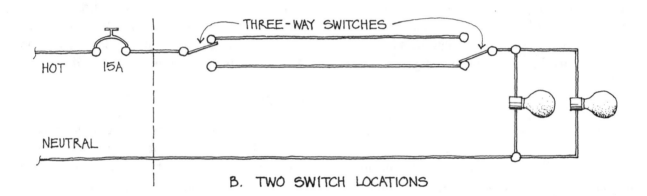

B. TWO SWITCH LOCATIONS

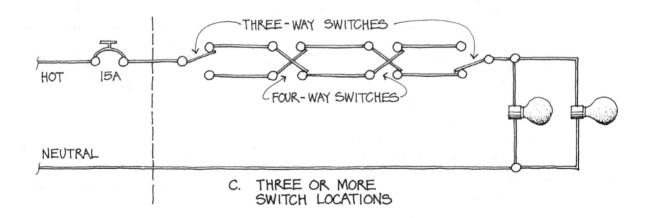

C. THREE OR MORE
SWITCH LOCATIONS

LIGHTING SWITCH CIRCUITS

Some circuits must be oversized. For example, the rating of a motor should not be greater than 80 percent of the capacity of the circuit that feeds it. Thus, any circuit that carries motors should have the rated load multiplied by 1.25 (1/0.80), to determine the wire size. The same factor should be applied to any circuit that is expected to operate continuously for three hours or more.

Conduit

Wires must be physically protected in addition to being insulated. In commercial applications this is accomplished by housing them in a conduit. The conduit size is designated by its interior diameter, and the number of wires of any given size that can fit into a given conduit is specified by code.

TABLE 6.2 – WIRE GAGES AND CONDUIT SIZES

AWG Size	Ampere Capacity 60 C Insulation	Conduit Size for Three Wires (Inches)
14	15	1/2
12	20	1/2
10	30	1/2
8	40	3/4
6	55	1
4	70	1
3	80	1 1/4
2	95	1 1/4
1	110	1 1/4
0	125	1 1/2
00	145	1 1/2
000	165	2
0000	195	2
250 MCM	215	2 1/2
300 MCM	240	2 1/2
350 MCM	260	2 1/2
400 MCM	280	3
500 MCM	320	3

There are several types or classes of conduit.

Rigid conduit is the safest conduit, and has the same wall thicknesses as Schedule 40 plumbing pipe. All connections are rigid and threaded, similar to plumbing pipe. The rigid conduit is installed, and then any number of wires up to its rated capacity may be pulled through it. In exterior applications, the conduit must be galvanized, while for interior situations it may simply be enamel coated.

Intermediate metallic conduit (IMC) is a steel conduit with thinner walls than plumbing pipe, slightly less expensive, and generally as acceptable for conduit as rigid conduit.

Electrical metallic tubing (EMT) is the thinnest of the simple metal conduits. It is galvanized, and connections are made with a special clamping system rather than threading. It is sometimes known as *thin wall*.

Flexible metal conduit is available both with and without a flexible waterproof jacket. It is commonly called "flex" or occasionally referred to by the product name "Greenfield." It can be used everywhere except underground.

A conduit similar to "flex" is *interlocked armored cable,* which consists of a prewrapped set of wires encased in an interlocking metal spiral armor. It is factory assembled, and no wires may be added in the field. It is designated *BX* cable, and it may not be used underground or embedded in concrete. No wires may be pulled through it in the field or after it is installed.

In office construction, where there may be several types of cable and power services, and the office layouts may change, there are special *power grid* floors or *cellular metal* floors. The concrete floor is poured directly over the floor system, which has knockout panels at regular intervals to allow access to the different raceways.

An alternative to conduit in residential construction is *sheathed* wire or "Romex." This consists of two insulated live wires and one ground wire, all encased in a plastic sheath. It is officially designated as type *NM* or *NMC* cable. It is strung inside walls, and sometimes in exposed areas such as garages. No other covering is required. It may not be used in commercial garages, and cannot be embedded in concrete.

CALCULATIONS

Most calculations are beyond the scope of this course, and not required of the architect or candidate.

However, you should be aware that the voltage drop due to the resistances of the wire in a given circuit may become noticeable in a large circuit. No more than 3 percent is allowed in lighting circuits, and no more than 5 percent in circuits supporting motors.

Load Estimation

It may be necessary for the architect to estimate the overall electrical load early in a project. This can be done by estimating the wattage per square foot, based on general experience for various building functions.

There are also minimum wattage per square foot requirements to provide for lighting in buildings. Of course, the actual load should be determined by the lighting calculations, not by estimates of wattage per square foot. (See Lesson Seven.)

TABLE 6.3 – LOAD ESTIMATING FACTORS			
	Watts/Square Foot		
Load	**Low**	**Avg.**	**High**
Lighting	2	3	5
Convenience Outlets	1	2	3
HVAC*	4	5.5	7
Miscellaneous	0.2	0.5	0.7
*Fuel-fired heating, electrically driven refrigeration.			

SAFETY CONSIDERATIONS

Short Circuits

Strictly speaking, a short circuit occurs when two conductors that are adjacent to each other lose so much insulation that a current flows directly between them. Since there is little resistance, a very high current may result, which can cause the wiring to get quite hot. Combustion within the walls may take place. This is especially dangerous because it may smolder and spread for some time before it is detected.

In practice, the term short circuit is applied to almost any situation where the current is flowing where it shouldn't. Three types of protection are available.

Shutoff Devices

Fuses are devices that are composed of a soft metal link in a glass plug or fiber cartridge, which are rated at a certain current flow. If the current exceeds that rate, the metal link will get hot enough to melt, breaking the circuit. Obviously, fuses are used only once, and must subsequently be replaced. The largest glass plug fuse is rated at 30 amps, but cartridge fuses are available at much higher ratings.

Circuit breakers are devices that automatically disconnect a circuit when the current is excessive. The circuit breaker, however, may be reset after the trouble has been found and corrected. Circuit breakers may be used as a backup switch to shut off an area or circuit being worked on or examined. Although they are more expensive than fuses, circuit breakers are employed in nearly all commercial applications, since they are simple, require no replacement, and therefore have low maintenance costs.

The third type of protection is called a *ground fault (circuit) interrupter (GFI or GFCI)*. It detects a continual current lost to ground, even after the power is shut off. This current might not be great enough to cause a fire, and thus might not trip the circuit breaker or melt the fuse, but it is undesirable nonetheless. After detecting such a current, the GFI breaks the circuit. Such devices are required on any circuit of 15 or 20 amps that serves a bathroom, garage, or outdoor area, as well as temporary circuits on construction sites. All large high voltage circuits, such as 480/277 volt 1,000 amps, are required to have ground fault protection.

Grounding

Grounding is a basic safety precaution. A ground wire is fastened to an element that provides a path directly to the ground, thus dissipating any electric current with little or no resistance, and averting possible damage or injury. Thus, many appliances are housed in a metal casing, and the metal casing is grounded. If there is a short circuit inside the case, the current will pass through the case to the ground wire and be harmlessly dissipated rather than passing through the case to an individual who might touch it, resulting in possible injury.

Ground wires are covered with green insulation, or may even be bare. They are typically fastened at some point to a steel cold water pipe in the plumbing system, which should provide a direct path into and under the ground, where the current will be dissipated. Three-pronged outlets have the third prong connected to a ground wire.

SERVICES

All of the services arriving on the site are called the *service drop,* and consist of the wires from the main line, a transformer, a meter and a disconnect switch.

To avoid excessive voltage drop and flicker, the distance from the transformer to the meter should not exceed 150 feet. The minimum service for a residence is now *100 amps.* In residences, the panel and disconnect are usually located outside the building where they are accessible to firefighters. In commercial construction, they may be located inside the building, but must be directly accessible from an exterior door.

Meters

The electric usage in a building may be measured in two ways. In residential applications, only the total consumption is measured. The unit of measure is the watt hour, which again is usually in thousands, called kilowatt hours (kwh). Costs range from 8 to 18 cents per kwh.

In larger buildings, not only is the total consumption rate measured, but the *peak demand* as well. This is because large peaks require the utility company to build more power generating capacity to meet the peak, which will then sit there, idle, for the remainder of the time. This is extremely inefficient and expensive. The charge associated with the peak demand is called a *demand surcharge.*

Emergency Power Sources

Emergency power is required for lighting exit passages and exit signs, as discussed in Lesson Seven. In addition, many institutions such as hospitals require life support equipment and operating room equipment to be safe from power interruption. In many cases, elevators and other equipment may also require backup power.

For lights, backup power is often provided by batteries that are continuously recharged while power is on, and which take over when power is lost. Since batteries are typically 12 volt, fluorescent lamps requires some sort of power conversion.

For larger equipment, a diesel generator with an automatic starting switch and an automatic transfer switch is usually provided in the equipment room. This generator should have at least a *two hour* supply of fuel in reserve.

BUILDING AUTOMATION

Building controls continue to become more and more complex. This is most evident in HVAC and elevator controls, but other building functions are coming under semi-intelligent or computer control. Loads may be shifted to a different time of day to avoid peak demand charges. Lighting may be controlled by time clock or photocell. Fire equipment may be automatically activated and controlled, closing fire doors throughout the building and actuating sprinklers. Emergency intercoms are often provided in high rise construction. Automatic security controls are not only numerous, but sometimes rather creative.

All of this equipment consumes electric power, or is electrically activated, and will continue to increase electrical loading and complexity in buildings as more functions come under computer control, until we begin to have "intelligent" buildings.

SUMMARY

We have learned the basic physics and terms of electricity, and many of the applications in buildings. We have reviewed direct and alternating, single- and three-phase current. We know how to calculate current through series and parallel resistances, and how power is generated. We have been made aware of some of the safety precautions necessary for this amazingly adaptable form of power. Electricity and electronics will continue to affect our lives and our buildings, and understanding the material covered in this lesson will help prepare the candidate for the examination, as well as for actual practice.

LESSON 6 QUIZ

1. Given a 120 volt outlet and a hair dryer with a resistance of 8 ohms (Ω), what is the current flow through the hair dryer when it is turned on?

   A. 15 amps   **C.** 1.5 amps

   B. 6.7 amps   **D.** .67 amps

2. What is the wattage rating of the hair dryer in the previous question if the power factor is 1?

   A. 1,800 watts   **C.** 18 watts

   B. 32 kilowatts   **D.** 32 watts

3. Given three *parallel* paths, two paths with a resistance of 4 ohms (Ω) and one with a resistance of 2 ohms (Ω) what is the total net resistance?

   A. .125 ohms   **C.** 1.25 ohms

   B. 10 ohms   **D.** 1.0 ohms

4. If the voltage between the neutral and the peak of a three-phase current is 120 volts, what is the voltage between two phases?

   A. 120 volts   **C.** 240 volts

   B. 208 volts   **D.** 277 volts

5. A three-wire transformer means that

   A. the primary winding is one of three input options.

   B. the transformer generates three-phase current.

   C. the secondary winding allows two different voltage takeoffs.

   D. there is a tertiary winding.

6. To provide five light switches for a large room, we need

   I. two-way switches.

   II. three-way switches.

   III. four-way switches.

   A. I and II   **C.** I and III

   B. II and III   **D.** I, II, and III

7. A variable speed mixer motor is most likely a

   A. universal motor.

   B. DC motor.

   C. three-phase induction motor.

   D. synchronous AC motor.

8. Which conduit system is unacceptable for embedding in concrete?

   A. Rigid conduit

   B. Flex conduit

   C. IMC conduit

   D. Armored cable

9. Which of the following is true about a ground fault interrupter?

   I. It is a form of protection against short circuits.

   II. It may be actuated even when there is no equipment switched on in the circuit.

   III. It can completely disconnect a circuit.

   A. I and II   **C.** I and III

   B. II and III   **D.** I, II, and III

10. What is a demand surcharge?

  A. A negative charge which builds up in a circuit when the current is too great.

  B. The copper side of a capacitor.

  C. A method by which utilities attempt to reduce the need for new power plants.

  D. The factor that triggers a circuit breaker.

LIGHTING

INTRODUCTION

We have discussed the thermal processes in buildings and the associated systems. The processes that deal with light are similar in that they are natural processes, yet different in that our eyes are much more sensitive than our thermal senses, and the visible spectrum is a very narrow part of the radiation spectrum.

LIGHT AS THE DEFINER OF ARCHITECTURE

Architects use light in two ways. First, lighting, whether natural or artificial, allows us to see so that we can perform our tasks. In that sense, lighting makes a space usable. But in addition, forms and spaces themselves are perceived in terms of light. How we feel about a building, whether the concept is communicated or the sculptural nature of the building is appreciated, is also determined by light. All of the great architects understood this. Aalvar Aalto, Louis Kahn, and Le Corbusier spoke of light in nearly reverent terms, and made it a key factor in their buildings.

Perception and the Eye

We define light as that part of the electromagnetic radiation spectrum that can be perceived by the human eye. This ranges from blue light, at wavelengths around 450 to 475 nanometers (a nanometer is one millionth of a millimeter), through green and yellow light (at 525 and 575), to red light (at 650). White light is the combination of all of the wavelengths. When we see a wall surface as blue, what really happens is that the white light shines on the wall, and all of the wavelengths except the blue are absorbed by the wall. The blue wavelength bounces back, and is sensed by the eye.

The eye is composed of a focusing device, the lens; a device that controls the amount or brightness of the light admitted to the eye, the iris; and a sensing surface called the retina. The retina is composed of two types of nerve pickups: the cones, which sense colors, and the rods, which sense black and white. The rods work efficiently at very low light levels, such as moonlight. The cones give more information, but require more light. In a very dark room, you lose the sense of color, even if you can still see well enough to move without bumping into things.

The eye is astoundingly adaptive. It can adjust from levels below 1 footcandle to levels over 10,000 footcandles in moments. It is only damaged when the change is too rapid, or when most of the background is dark, but one spot is intensely bright. Such extreme contrasts are known as *glare*.

Perception and the Mind

Incoming information from the eye is analyzed by the mind, which sorts it and interprets it. A good example is our depth perception. Because our eyes are separated, there is a slight difference between what each of them sees. Our brain compares the two images, from which it determines the distance to the object seen. The mind also sorts foreground from background using perspective clues and color clues. Parallel lines approach each other in the distance. Warmer and

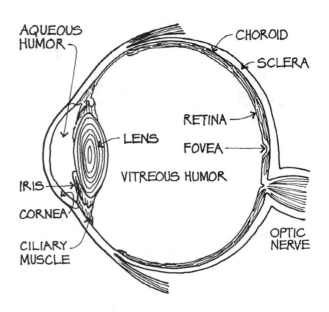

THE EYE

brighter colors tend to be interpreted as nearer, while cooler and darker colors are interpreted as more distant. Shadow information is processed to conclude the shape and form of the objects viewed. Our mind gives us a three-dimensional interpretation of what is really only two-dimensional information arriving at the eye. Sometimes the brain misinterprets this information and may confuse foreground and background, pattern and objects, or it may be fooled, as in optical illusions.

CONCEPTS AND TERMS

Several concepts and terms that are commonly used must be defined.

Transmission, Reflection, Refraction, and Absorption

All light that strikes a surface is either transmitted, reflected, or absorbed. *Transmitted* light passes through the material. If the *image* is transmitted, the material is called *transparent*. However, material that is transparent may

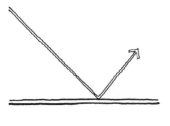
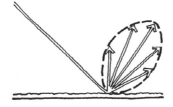
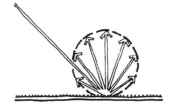

POLISHED SURFACE   ROUGH SURFACE   MATTE SURFACE
SPECULAR   SPREAD   DIFFUSE

DIFFUSE AND SPECULAR REFLECTION

change the image, such as the lens on a pair of glasses. This is called *refraction*, and occurs to some extent with nearly all transparent materials. When one looks at a fishing line in clear water, the line seems to bend sharply just below the surface. It is not the line that is bent, but the path of the light rays from the line, through the water and air interface.

If no image at all is transmitted, but there is still light passing through the material, it is called *translucent*. Frosted glass is an example. If the light is bounced off the material, it is called *reflective*. If the reflected image is maintained, as with a mirror, the surface is called *specular*. If the image is not maintained, as with a matte white finish, the surface is called *diffusing*. If absolutely no light passes through the material, it is called *opaque*. All of the light is either reflected or absorbed, or both.

Direct and Diffuse Light

Light is usually available to us in two forms. *Ambient* or *diffuse* light is the kind of light experienced on an overcast day. There are no sharp shadows, because the light is coming from all directions. In a building, this is analogous to a luminous ceiling, or a white ceiling lit by coves around the sides. This lights the entire room or area, and is therefore referred to as *area* lighting.

Direct light is the kind of light that comes directly from the sun on a sunny day. There are very sharp shadows, and the light is very strong. There are also very distinct reflections from shiny objects. Direct light is analogous to the light from a projector, or to a lesser extent, from a drafting lamp. It is most often useful when aimed at a task requiring special attention, and when so used is called *task* lighting.

Flat surfaces, such as murals, paintings, and papers or books, are best viewed in diffuse light, which prevents *veiling reflections* or *reflected glare*. Strongly modeled objects, such as sculptures, are more amenable to dramatic lighting, such as direct lighting, which casts sharp shadows, allowing us to understand the form.

Kelvin and Color Rendition Index

"Perfect" white light consists of a complete spectrum of wavelengths, with an even distribution. However, white light that has been transmitted through a translucent surface or reflected off of a surface is often shifted in color, or missing some part of the spectrum. Similarly, "artificial" light created by bulbs, tubes, or lamps sometimes has parts of the spectrum missing, or may have the distribution shifted one way or another. The measure of how well light actually shows true color is called the *color rendition index (CRI)*. This term is most often

Actual Size

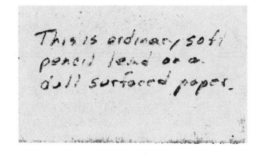

Top Lighting

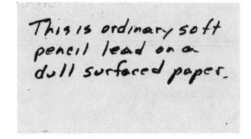

Side Lighting

Five Times Size

Top Lighting

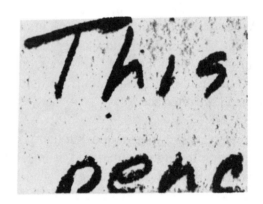

Side Lighting

At this size, you can see the actual reflections that break up the image

VEILING REFLECTIONS

Used by permission of Pilkington North America, Inc.

employed with artificial lighting. The best possible rating is 100, in which there are no colors missing. Another way of rating white light is called the *color temperature*, which comes from the theoretical relationship between the temperature of an object and the color of the light emitted, e.g., red hot to white hot to daylight (the surface of the sun is at approximately 6,000° Kelvin). The filaments or phosphors in the light source are not necessarily at the temperature indicated, but the color of the light is described nonetheless. We will discuss this in more detail when we cover artificial lighting.

BASIC PHYSICS

In everyday English, we use many lighting terms interchangeably and often imprecisely. In dealing with lighting in buildings, however, our terminology must be precise, and therefore candidates should become familiar with the definitions that follow.

Power or Intensity

The amount of light put out by a source is called the *intensity (I)*. The unit of measure for the intensity of a source is the amount of light coming from a single candle, called one *candlepower (cp)*.

Flux

If we drew a one foot square in midair, at a distance of one foot from a one candlepower source, the amount of light flowing through that square would be termed one *lumen (I)*. This flow through a theoretical surface is called the *flux (F)*.

Illumination

If we were to place a candle one foot from a blackboard there would be one lumen arriving on one square foot of blackboard surface. This is the *illumination (E,* since we already used I), and its value would be called one *footcandle* (fc). $E = F/A$.

Luminance

Last of all, the amount of light leaving the surface of the blackboard would depend on its reflectivity and would give us a measure of how bright it looked, or its *luminance.* For example, a perfectly reflective surface exposed to an illumination of one footcandle would have a luminance of one *footLambert (fL).* Brightness, or luminance, may also refer to the amount of light passing *through* a translucent surface. For example, a white surface with a reflectance of 80 percent and a white material with a transmittance of 80 percent will both have the same brightness if exposed to the same illumination.

The translucent surface will simply have that brightness on the far side, rather than the near side.

Inverse Square Law

Several basic rules may be used in lighting calculations. If the source of light may be approximated as a point (a candle, a light bulb, even a single tube or fluorescent fixture), the flux, and resultant illumination, is inversely proportional to the square of the distance from the surface.

For example, if a lamp has an intensity of 1,600 cp, a perpendicular surface 10 feet away has an illumination of:

$$E = I/d^2 = 1,600 \text{ fc}/(10 \text{ ft})^2$$
$$= 16 \text{ fc}$$

If we double the distance to the surface, to 20 feet, the illumination will be:

$$E = I/d^2 = 1,600 \text{ fc}/(20 \text{ ft})^2$$
$$= 4 \text{ fc}$$

Thus, doubling the distance cuts the illumination to one quarter. This can also be expressed by the relative formula:

$$E_2 = E_1 (d_1/d_2)^2$$

Thus the same example becomes:

$$E_2 = E_1 (d_1/d_2)^2 = 16 \text{ fc } (10 \text{ ft}/20 \text{ ft})^2$$
$$= 4 \text{ fc}$$

We can examine the brightness of surfaces in the example above. If we place the lamp 10 feet from a white wall with a surface reflectance of .75, the brightness or luminance of the wall is

$$L = 16 \text{ fc} \times 0.75 = 12 \text{ fL}$$

If we place a piece of frosted glass with a transmittance of .75 at 10 feet from the lamp, the brightness or luminance is:

$$L = 16 \text{ fc} \times 0.75 = 12 \text{ fL}$$

LIGHTING SYSTEMS

Different artificial sources produce different kinds of light, and vary significantly in their efficiency (or efficacy, which is the calculated lumen output per watt input). There are three general categories: incandescent, fluorescent, and high intensity discharge.

Incandescent

An incandescent lamp or bulb contains a *filament,* usually a tungsten alloy, which is heated by passing an electric current through it. It glows, giving off light and a significant amount of heat. The gas in the lamp is inert, such as nitrogen or argon, and does not interact with the filament or corrode it. Incandescent light is typically "warmer" than sunlight or daylight, rich in yellows and reds, and weak in greens and blues. Because much of the energy is wasted in the production of heat, incandescent lighting is the least efficient type of artificial light. It is also characterized by a short lifetime for individual bulbs. An output of *15 to 18 lumens per watt* is common, and a lifetime of around 2,000 hours is typical. The lifetime and output of bulbs are inversely related. Thus, burning a lamp at a lower voltage results in less light and a "warmer" color, but increases the lifespan of the lamp.

Incandescent lamps come in various shapes, with different characteristics. The most common shape is known as the A shape, which is the kind found in table lamps. Incandescent bulbs are sized in terms of wattage, and in multiples of 1/8 inch in diameter. For example, a 100 watt A-19 bulb uses 100 watts and is 19 × 1/8 inch, or 2.375 inches in diameter. R and PAR lamps have an internal reflector so that all of the light comes out of the front of the lamp. These are both more expensive and more effective than an A lamp at lighting a specific object. There are lamps with and without diffusing surfaces. There are lamps that run at lower voltages, such as 12 or 24 volts, which allows a smaller filament and results in a better focus of the beam.

Tungsten-halogen lamps are incandescent lamps that house the filament within an inner quartz envelope that can tolerate higher operating temperatures. This also contains a special halogen gas that prevents the evaporated metal from the filament from depositing on the inner surface of the quartz. Thus, the filament can be run at much higher temperatures, which produces more light and a slightly better color. In addition, the redeposition of the metal back onto the filament extends the life of the lamp slightly.

Fluorescent

Fluorescent lighting is a much more efficient system based on passing a current through gases inside a glass tube. This releases energy in the form of free electrons and gas ions. The glass tube can be lined with phosphors which in turn are excited by the ions, and which then glow in characteristic colors. If the phosphors are combined correctly, good color renditions can be achieved.

Unfortunately, it is impossible to get the current to arc through the gas at 110 volts, and therefore a transformer is necessary. In addition, once the arc has been formed, the resistance of the tube changes, and the circuit must be adjusted to avoid excessive current. A fluorescent fixture actually consists of the lamp and an associated

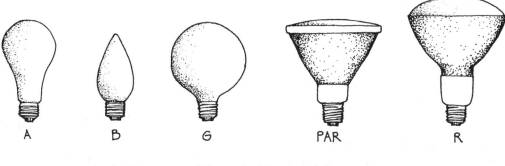

INCANDESCENT LAMP SHAPES

ballast that controls the voltage and the current to the lamp. Ballasts are sometimes noisy, and are assigned a Sound Rating from A to E, A being the quietest.

Four foot lamp lengths utilizing 40 watts or less are most common, although there are now small U-shaped lamps five inches in length, as well as large high-output lamps up to eight feet in length. There are a myriad of color combinations available. *Cool white* produces the most lumens per watt, but its color is unflattering to most people's skin tones. *Warm white* is somewhat better. *Cool white deluxe, warm white deluxe, royal white*, and the *SP* series lamps provide better CRI at slightly greater costs, because of the rare earth phosphors required. The lifetime of a fluorescent tube is determined not only by the number of hours it is lit, but also by the number of times it is switched on and off, which tends to wear it out. Based on a three-hour burning period every time the lamp is turned on, the average lifetime is about 10,000 hours. Efficiencies usually range from *60 to 80 lumens per watt.*

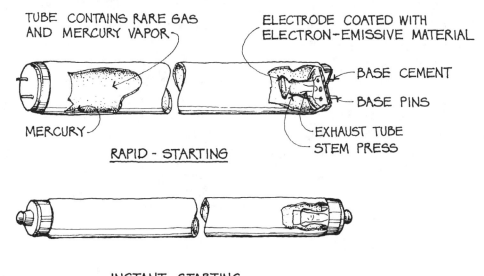

FLUORESCENT TUBES

High Intensity Discharge

High intensity discharge (HID) lamps consist of a lamp within a lamp, which is run at a very high voltage. There are four general types, only three of which are used in architectural lighting. The first HID lamps were *mercury vapor* lamps. They produce a very bright, clear, bluish light that unfortunately makes humans look unhealthy. This can be improved by adding phosphors to the inner surface of the outer envelope. Such lamps are called *mercury vapor deluxe*. Lifetimes are in the 24,000 hour range, and output is up to *50 lumens per watt*.

The next improvement in HID lamps was the inclusion of a *metal halide* gas, typically iodine, in the inner envelope, which shifted the color and improved the efficacy to approximately *80 lumens per watt*. However, the expected lifetime for the lamp dropped to about 10,000 hours.

The most efficient of the architectural HID lamps is the *high pressure sodium (HPS)* lamp. It develops up to *110 lumens per watt,* and has a 24,000 hour life expectancy. Unfortunately, the color rendition suffers somewhat, although new coated "deluxe" models seem to be a vast improvement at a very slight drop in efficacy.

Low pressure sodium lamps have the highest ratings in both lifetime and efficacy, but produce a monochromatic yellow light suitable solely for security lighting, since there is no color rendition at all. Everything is seen as if it were black and white, only it is black and yellow instead. Red looks black, blue looks black, and white and yellow look identically yellow.

The least efficient lamp type is normal incandescent. Tungsten-halogen is a slight improve-

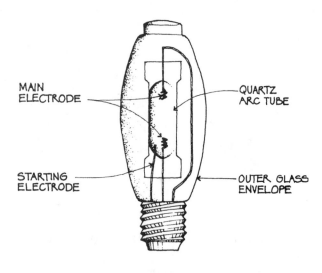

HID LAMP

ment, followed roughly by mercury vapor, then fluorescent and metal halide in a tie, and finally the most efficient, high pressure sodium. Low pressure sodium is even more efficient, but can only be used in security lighting or similar applications.

ARTIFICIAL LIGHTING CALCULATIONS

There are two common methods for calculating light levels. One works best for a single fixture, or small numbers of fixtures, and is called the *point grid* method. It takes very careful account of orientation and distance, but ignores surrounding reflection. The *zonal cavity* or *room cavity* or *lumen* method is based on a uniform distribution of a large number of fixtures, and takes into account the reflectivity of the ceiling and walls, and the comparative volumes of the top, middle, and bottom of the room. Both will be covered briefly, since detailed calculations are not likely on the examination. The Illuminating Engineering Society (IES) publishes an excellent reference manual for use when detailed calculations are necessary.

Point Grid Method

The point grid method is based on the formula:

$$E = I \cos \theta / d^2$$

where

  E = illumination at the receiving surface

  I = intensity at the source when viewed from the direction of the receiving surface

  θ = the angle between a perpendicular (normal vector) to the receiving surface and a line from the source to the surface

  d = the distance from the source to the surface

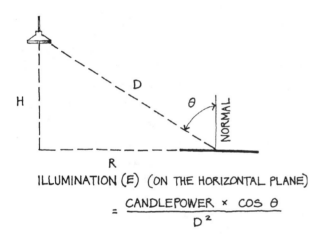

ILLUMINATION (E) (ON THE HORIZONTAL PLANE)

$$= \frac{\text{CANDLEPOWER} \times \cos \theta}{D^2}$$

POINT GRID METHOD

The intensity in a given direction is taken from polar plots of the fixture intensity sometimes known as *candlepower distribution curves*. These graphs show how much light is given off at any given angle from a vertical reference line. If all of the light goes up, the fixture is called an *indirect* fixture, because the light will reach the work plane only after bouncing off the ceiling. This is particularly suitable for uniform light in rooms with computer terminals. If all of the light goes down, it is called a *direct* fixture.

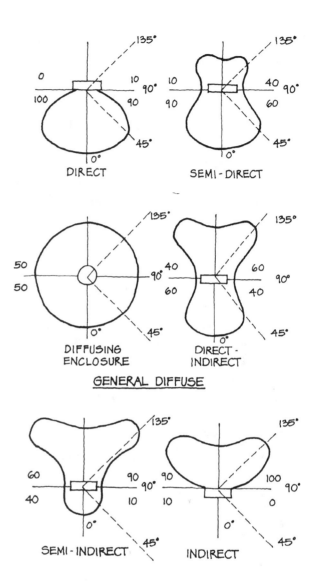

CANDLEPOWER DISTRIBUTION CURVES

If the light spreads widely to either side, it gives a uniform illumination on the floor, but may cause glare by shining into the user's eyes even when he or she is not looking up. If the beam spread is narrow, the fixtures must be closely spaced to avoid spotty illumination on the floor and work plane.

Abney's Law simply states that the light arriving at a surface is the sum of the light arriving from all of the sources, and can be expressed by repeating the point grid formula for each source:

$$E = I_1 \cos\theta_1/d_1^2 + I_2 \cos\theta_2/d_2^2 \\ +\ldots+ I_n\cos\theta_n/d_n^2$$

This means that we can add the light from all of the fixtures in a room, one at a time, and get the correct result for all of the direct light, but the reflected light is not considered.

Zonal Cavity Method

The method most commonly used for office, commercial and factory spaces is the zonal cavity method. It is based on a coefficient of utilization (CU) for each fixture type that takes into account the direction that the fixture throws its light, the reflectances of the ceiling cavity, the middle level of the walls, and the zone between the work surface and the floor. The CU theoretically varies between 0 and 1.0, with most values in the .5 to .8 range. The formula also considers maintenance factors and dirt, and is expressed by the equation:

$$E = (N \times n \times LL \times LLD \times DDF \times CU)/A$$

where
E = the illumination in footcandles
N = number of fixtures
n = the number of lamps per fixture
LL = number of lumens produced
  per lamp
LLD = the lamp lumen depreciation factor
  (which accounts for the effects of
  aging on lamp output)
DDF = the dirt depreciation factor for the
  fixture based on scheduled maintenance and how dirty the surroundings
  are
CU = the coefficient of utilization, calculated as discussed in the paragraph
  above

A = the area of the working plane (or
  floor) that will be illuminated by the
  fixtures

The equation can be manipulated so that if the designer knows the desired illumination level in footcandles (E), he/she can solve for the necessary number of fixtures (N). The result is

$$N = E \times A/n \times LL \times LLD \times DDF \times CU$$

RECOMMENDED ILLUMINATION

The amount of light required to see well varies greatly with the age of the observer and the task at hand, and there is significant disagreement about the ideal illumination levels. How diffuse the light is also affects these levels. The optimum lighting is called the *equivalent spherical illumination (ESI)*. It is based on a theoretical sphere surrounding the object being illuminated with the light cast evenly from all parts of the sphere, eliminating any shadows and any reflected bright spots. 100 footcandles of light from a single lamp may yield a level of legibility equivalent to only 50 footcandles ESI. The concept is important, but the calculations are beyond the scope of this course.

DAYLIGHT CALCULATIONS

Perhaps the most significant rediscovery of the past ten years is the use of natural daylighting to save energy and improve the quality of light.

Whenever we can turn off a fixture, we reduce the electrical load and reduce the heat generated within the space as well. In hot climates this results in a double benefit. Even in cold climates, there is often an economic benefit. In addition, daylight is a diffuse light source, and the color rendition is perfect.

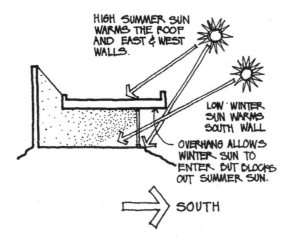

HIGH SUMMER SUN WARMS THE ROOF AND EAST & WEST WALLS.

LOW WINTER SUN WARMS SOUTH WALL

OVERHANG ALLOWS WINTER SUN TO ENTER BUT BLOCKS OUT SUMMER SUN.

SOUTH

THE SUN'S LOCATION VARIES WITH THE SEASONS

However, certain precautions should be taken. Direct sunlight should be shielded, diffused by diffusing glass, or bounced off of diffusing surfaces. Horizontal skylights should be introduced with care, because of the potential for great heat gain in the summer and little heat gain in the winter. (Remember that horizontal surfaces get *more* sunlight in summer and *less* sunlight in winter.)

Artificial lighting should be provided even in daylit areas, for nighttime illumination. Lighting fixtures may be connected to switches in such a manner that fixtures can be turned off in areas where daylight is sufficient, while other fixtures may be left on. Even better, dimmer controls may be connected to a photosensitive device that dims the lights successively as the daylight level increases. Both stepped and continuously dimmable switches and ballasts are now available for fluorescent fixtures.

Daylighting Strategies

There are several architectural elements that can be used in daylighting strategies.

A *light shelf* is an overhang with glass above it, which reflects light into the room and up onto the ceiling. Such light shelves are usually placed above head height, to avoid reflecting glare into the eyes of the occupants. A *glass transom* is a transparent area over a door, or even over shelves, bookcases, etc., which lets light pass from one room into another, while providing some security and acoustical privacy. A *sawtooth* is a roof with a series of vertical or nearly vertical glass surfaces, usually facing north. This lets in the even, diffuse light, but limits the direct sunlight to early summer mornings and late summer evenings. For more information on daylighting techniques, check some of the references in the bibliography.

There are two methods for calculating the amount of daylight in a space based on orientation, windows, and internal and external reflection.

Lighting and Sustainable Design

The illumination of the interior of a sustainably designed building requires a holistic approach that balances the use of artificial and natural lighting sources.

1. *Daylighting*

   Properly filtered and controlled solar radiation may provide a valuable source of illumination to a building interior. This process is called "daylighting" (simply having properly designed fenestration that allows natural sunlight to replace or dramatically reduce the need for artificial lighting).

   Because unwanted sunlight (particularly in summer months) can also add to the internal heat load of a building, the architect must be careful to allow only beneficial sunlight and reduce unwanted solar heat gain. There are several techniques for controlling daylighting:

   - Overhangs, fins, and other architectural shading devices.

   - Sawtooth (not bubble) skylight design, which allows the glass to face north for illumination—not south for solar heat gain.

   - Interior window shading devices, which allow solar gain during cool months, and the blocking of solar radiation during the warmer seasons.

   - Light shelves, which permit the daylight to reflect off the ceiling and penetrate farther into the interior without affecting views outside.

2. *Higher Efficiency Light Fixtures*

   In addition to a daylighting strategy, light fixtures that are more efficiently designed reduce energy cost and increase comfort, such as:

   - Fixtures that use fluorescent or HID lamps, which provide more illumination per watt than incandescent lighting.

   - Fixtures that are designed to diffuse or bounce the illumination off the ceilings or internal reflectors, which are more efficient, cause less glare (especially in an environment with computer monitors); and save operating costs.

   - Fixtures that have higher efficiency (T-8) fluorescent bulbs, which produce more lumens per watt and thereby diminish the heat generated by lighting.

   - Fixtures that offer dimming or multiple switching capability, which permit the architect a more energy efficient lighting design.

     Dimming or multiple switching fixtures allow the architect to design lighting patterns that blend nicely with daylighting opportunities.

     - For example, an office with perimeter fenestration allows daylighting supplemented with overhead lighting that can be dimmed or reduced.

     - The interior spaces, which are too far from the perimeter for daylighting, may be controlled with switches or dimmers that allow relatively higher levels of illumination.

     - The result is an even illumination pattern, which saves on artificial lighting costs, by relying on daylighting at the perimeter.

   - Fixtures that use higher efficiency lamps such as fluorescent, high intensity discharge (HID) sulfur lighting (exterior only).

   - Fluorescent fixtures that use high efficiency electronic ballasts.

     Additionally, the architect may avoid less efficient incandescent lighting where possible, install task lighting to supplement diffused ambient lighting, and install LED (light emitting diode) lighting for exit signs. LED lighting lasts longer than incandescent and is far less expensive to operate.

3. *Lighting Sensors and Monitors*

Where possible, lighting cost can be diminished by installing light monitors that sense occupancy conditions. As long as the room contains people, the lights will remain on. If people leave, the sensor will wait for a few minutes, then shut off all the lighting in the room.

Lighting sensors can be designed to operate with a preference for motion, heat (from people), or desired time of occupancy.

4. *Lighting Models*

Computer lighting models are one option that allows the architect to simulate the levels of sunlight that penetrate into a building design, depending on the building location, varying times of year, fenestration orientation, and design.

By incrementally altering fenestration (skylights, windows, or light transport systems) and the artificial lighting system, the architect may optimize the daylighting and artificial lighting systems for the building.

Lumen Method

The method developed in the United States is called the lumen or flux method. It is approved by the IES and was long promoted by Libbey-Owens-Ford as a public service. It is well suited to both clear skies and partly cloudy skies. The amount of daylight is calculated in only three locations in the room: five feet from the window, the middle of the room, and five feet from the back of the room. It can be used to calculate the amount of daylight from one window wall, or two window walls on opposite sides, but not from a corner window (windows on two adjacent walls).

Daylight Factor Method

The other method is called the daylight factor method, which was developed in Europe and assumes overcast or diffuse sky conditions. This method is sanctioned by the Commission Internationale d'Eclairage (CIE) and is often used in computer programs. The major benefit is that it can be used to calculate the amount of daylight at any location in the room, including the effect of corner windows.

The result of a daylight factor calculation is a number that expresses the amount of light at a particular interior location as a percentage of the light which is available on an exterior horizontal surface. For example, a daylight factor of 3 at a corner of the room means that 3 percent of the light available outside would arrive on the workplane in that corner. Since outside light levels can be quite high, 3 percent might be sufficient. If there were 2,000 fc on the ground outside, then a DF of 3 would result in an illumination level of $2{,}000 \times 0.03 = 60$ fc.

EMERGENCY AND EXIT LIGHTING

Most building codes require special emergency lighting at exits and for certain critical functions. Emergency lighting may be provided by a separate generator in the building, or by battery packs and small lights which would provide enough illumination to evacuate the building. Most codes do not allow lead-acid battery packs because of the fumes emitted, although there are exceptions with small batteries or proper ventilation. *Nickel-cadmium* batteries are slightly more expensive, are rechargeable, and emit no noxious fumes. For fluorescent emergency lighting, some kind of *transformer* and *inverter* are necessary, since fluorescents do not run on 12 volt DC current.

Exit signs themselves must be illuminated by at least two sources: the general illumination and an internal or special illumination. Even with the lights out and the room full of smoke, the exit signs should be visible to escaping occupants.

SUMMARY

We have studied light, the definer of space and form, starting with some of the processes involved in perception, and then the terms and physics of lighting. Different types of artificial lighting have different characteristic CRIs and efficacies. There are different ways of calculating illumination, both for single point sources and for uniformly lit rooms, as well as for natural lighting in clear and overcast skies. All of these factors together yield an understanding of lighting within buildings, as necessary to the examinee and to the architect. For further study, the Illuminating Engineering Society's *1984 Reference Volume* is recommended.

LESSON 7 QUIZ

1. Which color represents the highest Kelvin temperature and the shortest wavelength?
   - **A.** Red
   - **B.** Yellow
   - **C.** Green
   - **D.** Blue

2. What is the name of the sensing surface of the eye?
   - **A.** The lens
   - **B.** The iris
   - **C.** The retina
   - **D.** The cilia

3. Which of the following statements about glare are true?
   - I. Glare is always a reflection.
   - II. Glare is always an extreme contrast in brightnesses.
   - III. Glare is always a very high illumination.
   - **A.** I only
   - **B.** II only
   - **C.** III only
   - **D.** I, II, and III

4. A material which transmits light without transmitting the images on the far side is called
   - **A.** transparent.
   - **B.** translucent.
   - **C.** opaque.
   - **D.** reflective.

5. Which of the following statements are correct about the term diffuse?
   - I. Ambient light is usually diffuse.
   - II. A mirror is a diffuse reflecting surface.
   - III. Printing or writing is easiest to read when viewed in diffuse light.
   - **A.** I and II
   - **B.** II and III
   - **C.** I and III
   - **D.** I, II, and III

6. What is the unit which measures the amount of light striking a surface (the illumination)?
   - **A.** Candlepower
   - **B.** Lumen
   - **C.** Footcandle
   - **D.** FootLambert

7. If a lamp of 2400 cp intensity results in an illumination of 24 fc at 10 feet, what is the illumination at 20 feet?
   - **A.** 6 fc
   - **B.** 8 fc
   - **C.** 10 fc
   - **D.** 12 fc

8. The major advantage to low voltage incandescent lamps is that the lamps
   - **A.** are much brighter for a given wattage.
   - **B.** last much longer.
   - **C.** focus and aim their light more accurately.
   - **D.** can run at higher temperatures.

9. What is the diameter of an R-40 bulb?
   - **A.** 2.5"
   - **B.** 3.2"
   - **C.** 4"
   - **D.** 5"

10. Which of the following are true about fluorescent lighting?
    - I. Fluorescent lighting is more efficient than incandescent lighting.
    - II. Fluorescent lighting can have different CRI ratings, depending on phosphors.
    - III. Fluorescent lighting does not run on 12 volt DC current.
    - **A.** I and II
    - **B.** II and III
    - **C.** I and III
    - **D.** I, II, and III

ACOUSTICS

8

INTRODUCTION

Acoustics is the science of sound. Sound is the sensing, by the ear, of compression waves in a fluid medium, usually air. It is a vital part of our everyday lives, and a strong influence in the design of all buildings. Sound isolation is necessary for privacy, and in many cases for specific tasks, as well. But some buildings also have special acoustical requirements, primarily for musical performances or for communication, which are concerned with clarity and amplification. Therefore, the architect should have a basic understanding of acoustics.

BASIC PHYSICS

In many ways, sound is similar to light. They are both transmitted by waves, and they both obey the inverse square law: the intensity is inversely proportional to the square of the distance from the source. The processes of transmission, reflection, and refraction can apply to both sound and light.

Unlike light, however, sound can only be transmitted through a medium, such as air. The velocity of sound through air depends on the barometric pressure and altitude. It can also be transmitted through water, through the ground, or through building structure and materials. It can be reflected off of surfaces, and even focused toward a point by the proper reflective shape. Sound may also be refracted (bent) around objects. Light tends to cast much sharper shadows than sound. A wall in an open field will cast a very sharp light shadow, but its sound shadow is indistinct; sounds from the other side of a free standing wall are often clearly audible.

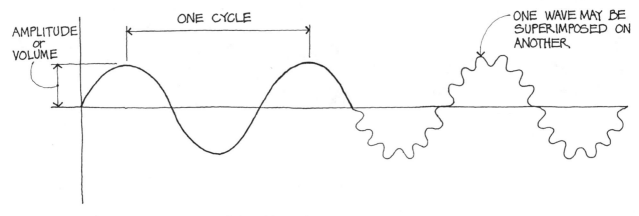

AMPLITUDE or VOLUME

ONE CYCLE

ONE WAVE MAY BE SUPERIMPOSED ON ANOTHER

SOUND WAVES

Just as we perceive the wavelength of light in terms of its color, we perceive the wavelength of sound in terms of its *pitch*. Each note of the musical scale represents a specific wavelength or frequency of sound. This is most easily seen when we look at a graph of the sound compression wave in air.

One complete wave form is called a cycle. The frequency of the sound (the pitch) is the number of cycles occurring per second. This may be abbreviated cps, but is more commonly known by the term *Hertz*. Thus, 60 cps = 60 Hz.

A sine wave is a "pure" tone, such as that produced by an electronic instrument. A sound may consist of one wave form superimposed on another. The ear can distinguish several notes at once, for example a chord, or even an entire symphony orchestra playing a single chord.

The ear can distinguish between a square wave at a particular pitch and a sine wave. The square wave sounds harsh, like a buzzer, or even an outboard motor. Different instruments, different voices, and other individually recognizable sound sources have different wave forms.

Although the human ear can hear sounds in the 20 Hz to 20,000 Hz range, it is most sensitive to sounds in the *125 Hz to 6,000 Hz* range. Many animals can hear much higher frequencies. Sounds below 20 Hz are often sensed as vibrations rather than sounds. Structure-borne sound in buildings is sometimes barely audible, but can still cause discomfort.

The height of the wave form is related to the amplitude or magnitude or intensity of the sound. Loud sounds have great amplitude, and represent a larger amount of energy stored in the wave, which may be measured in *watts/cm²*.

The human ear can respond to a tremendous variation in sound amplitudes without being damaged. The ratio between the amplitude of the quietest sound the human ear can hear and that of the loudest sound that does not cause pain is 1:10,000,000,000.

Logarithmic Scales

In order to deal with these variations in a manageable manner, acoustics uses logarithmic scales. The decimal logarithm (log) of a number is the exponent to which the number 10 must be raised to equal that number. For example, the log of 100 is 2, since $100 = 10^2$.

The five basic rules of logarithms are:

$$\log 10^n = n$$

$$\log A \times B = \log A + \log B$$

$$\log A/B = \log A - \log B$$

$$\log C^n = n \log C$$

$$\log 1 = 0$$

Sound Intensity Level

The basic unit of sound intensity level is called the *decibel* (after Alexander Graham Bell) and is expressed by the formula:

$$IL = 10 \log (I/I_0)$$

where

IL = intensity level expressed in decibels (dB)
I = intensity of the sound being measured
I_0 = reference intensity of 10^{-16} W/cm^2, which is the quietest sound that we can hear

Note that the *intensity* of sound is measured in power (watts) per square centimeter, but we generally deal with the *intensity level (IL)*, which is measured in decibels.

To get an idea of how useful the logarithmic scale is, let's solve for the intensity level of a sound that is 10,000,000,000 times as loud as the quietest sound we can hear.

$IL = 10 \log (I/I_0)$

$$= 10 \log \frac{10,000,000,000 \times 10^{-16}}{1 \times 10^{-16}}$$
$$= 10 \log (10,000,000,000/1)$$
$$= 10 \log 10^{10}$$
$$= 10 \times 10$$
$$= 100 \; dB$$

Obviously, 100 dB is a great deal easier to express than 10,000,000,000 $\times 10^{-16}$ watts/cm^2.

How does this work for small ratios, for example a sound that is twice the intensity of the reference intensity?

$$IL = 10 \log (I/I_0)$$
$$= 10 \log (2/1)$$
$$= 10 \times 0.301$$
$$= 3.01 \; dB$$

The number is still meaningful and easy to express.

Sound Power Level

We can also measure the power *at the source* and convert it to a logarithmic scale. This would measure watts instead of watts/cm^2 or watts/meter^2. The result would be:

$$PWL = 10 \log W/W_0$$

where
PWL = sound power level
W = power at the source measured in watts
W_0 = reference wattage, 10^{-12} watts

Sound Pressure Level

There is a third measure of sound, which is the pressure exerted by the sound wave on a surface at a given location. This is similar to intensity level, but not exactly, since it varies with barometric pressure as well. It is also expressed in logarithmic form:

$$SPL = 20 \log P/P_0$$

where
SPL = sound pressure level
P = pressure at the measured point, in newtons/meter^2
P_0 = reference pressure, 2×10^{-5} N/m^2

Of the three different types of sound measurement, IL is by far the most widely used. Notice the factor of 20 in the equation for SPL, which

TABLE 8.1 – TYPICAL IL LEVELS		
Sound	**SPL dB**	**Intensity Watts/Meter²**
Threshold of pain	130	10^1
Hard rock band	120	10^0
75 piece orchestra	110	10^{-1}
Loud auto horn 10 feet away	100	10^{-2}
Noisy urban street	90	10^{-3}
Truck passing by	80	10^{-4}
Automobile passing by	70	10^{-5}
Conversation	60	10^{-6}
Average office	50	10^{-7}
Quiet office	40	10^{-8}
Unoccupied office	30	10^{-9}
Whisper	20	10^{-10}
Rustle of leaves	10	10^{-11}
Threshold of acute hearing	0	10^{-12}

makes the numerical value of SPL approximately equal to that of IL. Thus, for the purposes of the exam, IL and SPL may be assumed to have the same value. PWL is different, and always represents the power *at the source.*

The problem with using a logarithmic scale is that when we have two sound sources, we cannot simply add their dB levels together. For example, two sources at 60 dB each result in a total IL of 63 dB, *not* 120 dB. Similarly, although doubling the distance from the source to the receiver cuts the intensity to one quarter (remember the inverse square law) this results in a change of only 6 dB in intensity level. Table 8.2 may be used to determine the decibel level that results when two sound sources are added together.

It is always best to use the logarithmic equations, which is not difficult once the five basic rules are understood, as shown in the following example:

Example #1

Given a source of sound ten feet away from the receiver, and an IL of 90 dB at the measured position (receiver),

1. What is the sound *intensity* at the receiver position?
2. What is the *sound power* of the *source*?
3. What is the sound *intensity level* at 80 feet from the source?

Solution:

1. The intensity level at the receiver is 90 dB. We can determine the intensity using the logarithmic formula

$$IL = 10 \log I/I_0 = 10 \log I/10^{-16}$$

$$90 = 10 \log I/10^{-16}$$

$$9 = \log I/10^{-16} = \log I - \log 10^{-16}$$

$$\log I = \log 10^{-16} + 9$$

$$I = 10^{-16} \times 10^9 = 10^{-7} \; watts/cm^2$$

TABLE 8.2 — ADDING TWO DB LEVELS	
Difference in dB Level	**Add to Larger dB Level**
0	3.0
1	2.5
2	2.1
3	1.8
4	1.5
5	1.2
6	1.0
7	0.8
8	0.6
9	0.5
10	0.4

2. The sound power of the source is related to the intensity at a given point by the equation

$$I = W/(4\pi d^2 930)$$

where

   I = intensity in W/cm^2
   W = power at the source in watts
   d = distance to the source in feet
   930 = conversion factor to translate between feet and watts/cm^2

   Transposing that equation to solve for W in terms of I, $W = I \times (4\pi d^2 930)$

   Substitute 10^{-7} watts for I, and 10 feet for d:

$$W = 10^{-7} \times (4\pi 10^2 930)$$
$$= 0.12 \text{ watts}$$

3. Doubling the distance from the source reduces the intensity to one quarter (the inverse square law) and reduces the IL by 6 dB. Since 80 feet is 8 times the distance of 10 feet, the IL may be calculated by successive reductions of 6 dB for each doubling of distance.

Thus, the IL at 20 feet is 6 dB less than at 10 feet, the IL at 40 feet is 6 dB less than at 20 feet, and the IL at 80 ft is 6 dB less than at 40 ft, resulting in a total drop of 18 dB at 80 feet, or 90 dB − 18 dB = *72 dB*.

Alternatively, we could use the inverse square law, followed by the logarithmic conversion.

$$I_2 = I_1 (d_1/d_2)^2$$

$$I_{80} = I_{10}(10/80)^2 = I_{10} \times 1/64$$
$$= (10^{-7} \text{ watts/cm}^2)64$$
$$= 1.5625 \times 10^{-9} \text{ watts/cm}^2$$

$$IL = 10 \log (1.5625 \times 10^{-9}/10^{-16})$$
$$= 10 \log 1.5625 \times 10^7$$
$$= 10 \times (7.19)$$
$$= 71.9 \text{ dB}$$

This is essentially the same result as we obtained before.

Weighted Scales for the Human Ear

The human ear is more sensitive to sounds in the middle frequencies than to those in very high or low frequencies. Therefore, weighted measuring scales have been developed. The scale that most closely represents the response of the human ear is called the A scale. When measurements using the A scale are converted to decibels, the resultant measure is designated dBA.

The human ear may also be temporarily affected by exposure to continuous loud noise levels. There can be as much as a 30 dB loss in sensitivity. Permanent damage is also possible. For this reason, the Occupational Safety and Health Administration (OSHA) has developed requirements intended to limit exposure to high noise levels in places of employment. Table 8.3 presents the permissible exposure limits for continuous sound levels.

TABLE 8.3 – OSHA PERMISSIBLE NOISE EXPOSURE	
Duration per day (in hours)	**Sound Level, dBA (slow response)**
8	90
6	92
4	95
3	97
2	100
1 1/2	102
1	105
1/2	110
less than 1/4	115

As with vision, the mind integrates the incoming sensory information from the ear. If one ear hears something a few thousandths of a second before the other, the mind infers the approximate direction of the source. The exception is when a source is either directly in front of, above, or directly behind the head, since the sound arrives at both ears simultaneously, and direction cannot be determined.

TRANSMISSION AND REFLECTION

All of the methods above refer to the behavior of sound in a free environment, such as an open field. However, there are some considerable distortions when we get into buildings. Just as with light, there is a great deal of acoustical reflection inside of a room. Near the source, the inverse square relationship holds, but as we get further from the source, the reflections dominate, and the distance relationship becomes less important, or even unimportant.

Similarly, one of our concerns within buildings is the transmission of sound from one room into another room through the separating wall. These two phenomena are both affected by sound absorption.

Sound Absorption

The reflection of sound in a room causes two things to occur. The noise level, or volume of sound, is greater than it would be for the same source in an empty field. There is a delay factor as well, so some sounds persist for a time in a very reflective space, even after the source

TABLE 8.4 – TYPICAL SOUND ABSORPTION COEFFICIENTS						
	125 Hz	**250 Hz**	**500 Hz**	**1,000 Hz**	**2,000 Hz**	**4,000 Hz**
Heavy carpet on concrete	.02	.06	.14	.37	.60	.65
Coarse concrete block	.36	.44	.31	.29	.39	.25
Heavy velour drape	.14	.35	.55	.72	.70	.65
Plate glass	.18	.06	.04	.03	.02	.02
1/2" gypsum board on 2 x 4 studs	.29	.10	.05	.04	.07	.09
Smooth finish plaster on lath	.14	.10	.06	.04	.04	.03
Water surface, at rest	.008	.008	.013	.015	.02	.025
Audience in upholstered seats, per square foot of seating area	.60	.74	.88	.96	.93	.85

has stopped. This is known as reverberation. Although reverberation is similar to an echo, in some respects it is different.

An echo is a discrete reflection of a sound, usually delayed 1/10th of a second or more. With a sufficient delay, an entire word may return separate and intact. A prime example is the echo from a hard canyon wall, or some other distant, fairly flat object. Reverberation is a more continuous reflection, over shorter time spans, such as an organ note dying out slowly in a cathedral, as the sound reverberates back and forth. Not only is sound magnitude related to absorption, but so is reverberation time.

There is an acoustical measure of reflectivity and absorptivity similar to that in radiation, designated by the Greek letter alpha (α). It is measured in a unit called a sabin, after Wallace Clement Sabin, a physicist who pioneered acoustical work.

Sabin lectured in a physics auditorium in which most lectures were unintelligible. He determined that this was because sounds took so long to die away that the syllables of one word would run together and even into the next word. He was able to reduce this phenomenon by putting horsehair pillows on the seats. The pillows absorbed the sound instead of reflecting it back, and there was a vast improvement in the students' ability to understand his lectures.

The absorptivity per square foot of any given surface varies from 0 (all sound is reflected) to 1.0 sabin (all sound is absorbed).

The absorptivity of a room is the sum of the different surface areas times their respective absorptivities, and is expressed by the equation:

$$A = S_1\alpha_1 + S_2\alpha_2 + S_3\alpha_3 + ... + S_n\alpha_n$$

A = total absorptivity

S_1 = surface area of material 1 in square feet

α_1 = absorptivity of material 1 in sabins

S_2 = surface area of material 2 in square feet

α_2 = absorptivity of material 2 in sabins

and so forth, for all the materials present in the room.

Materials may have different absorptivities at different frequencies, so a complete table will include the α value for each frequency and an overall *noise reduction coefficient (NRC)*, which is the arithmetic average of all of the different frequencies from 250 Hz to 2,000 Hz.

Example #2

Find the total absorptivity of an empty room at 1,000 Hz. The room is 20 × 30 feet in plan and 10 feet high. The floor is concrete covered with carpet, one 30 foot wall is plate glass, and the other three walls are half inch gypsum board over studs. (It is not unusual to disregard doors when calculating absorptivity, but they are most important when dealing with transmission.) The ceiling is of smooth finished plaster on lath.

Solution:

$$
\begin{aligned}
A = \quad & 2 \times 20 \times 10 \times .04 \text{ (two 20 foot walls)} \\
+ \ & 1 \times 30 \times 10 \times .04 \text{ (30 foot wall)} \\
+ \ & 1 \times 30 \times 10 \times .03 \text{ (30 foot window)} \\
+ \ & 1 \times 30 \times 20 \times .37 \text{ (carpet on floor)} \\
+ \ & 1 \times 30 \times 20 \times .04 \text{ (plaster ceiling)} \\
= \ & 16 + 12 + 9 + 222 + 24 \\
= \ & 283 \text{ sabins}
\end{aligned}
$$

most of which is represented by the carpet!

We can calculate the volume (amplitude) of the sound in an enclosed reverberant space using several equations. If we know the power in watts at the source we can calculate the intensity (watts/meter^2) from the equation

$$I = P/930A$$

where

I = intensity throughout the space (except very close to the source), in watts/meter^2

P = power at the source, in watts

A = total absorptivity of the space, in sabins

We can find the intensity level (IL) using the logarithmic equation on page 147. Once we know the IL in a room for a given condition, we can find the IL resulting from any given change from an easier formula:

$$NR = 10 \log A_2/A_1$$

where

NR = reduction in sound from case 1 to case 2, in decibels

A_2 = total absorptivity of the room in case 2, expressed in sabins

A_1 = total absorptivity of the room in case 1, expressed in sabins

Example #3

What would be the absorptivity if we provided drapes across the window in Example #2? What would be the reduction in sound?

Solution:

A =   2 × 20 × 10 × .04 (two 20 foot walls)
     + 1 × 30 × 10 × .04 (30 foot wall)
     + 1 × 30 × 10 × .72 (30 foot drapes)
     + 1 × 30 × 20 × .37 (carpet on floor)
     + 1 × 30 × 20 × .04 (plaster ceiling)
   = 16 + 12 + 216 + 222 + 24
   = 490 sabins

compared to 283 sabins.

The reduction in sound in that room would be

$$
\begin{aligned}
NR &= 10 \log A_2/A_1 = 10 \log 490/283 \\
&= 10 \log 1.73 \\
&= 10 \ (.238) \\
&= \textit{2.38 (dB)}
\end{aligned}
$$

We could still include the absorptivity of the glass wall, since it hasn't really been removed. It is not a significant proportion of the total. Hanging fabric tapestries or drapes in a concert hall provides one way of varying the absorptivity of the hall for different functions. The drapes may be retracted, or the tapestries removed, as necessary.

Reverberation

As we previously discussed, reverberation is the slow fading of a sound in an enclosed space. The amount of time that elapses before there is complete silence after a 60 dB sound has stopped is called the reverberation time (T_R). As spaces become larger, the reverberation time tends to increase. As absorptivities become greater, the reverberation time tends to decrease. This relationship is expressed by the formula:

$$T_R = .049 \ V/A$$

where

T_R = reverberation time in seconds
V = volume of the space in cubic feet
A = total absorptivity in sabins

Different functions have different optimum reverberation times. In Table 8.5, the optimum reverberation times for various functions are displayed in bar chart form.

Typically, speech should have a short reverberation time, to avoid the overlapping of syllables and resultant unintelligibility. Conversely, a pipe organ sounds best when there is a longer reverberation time.

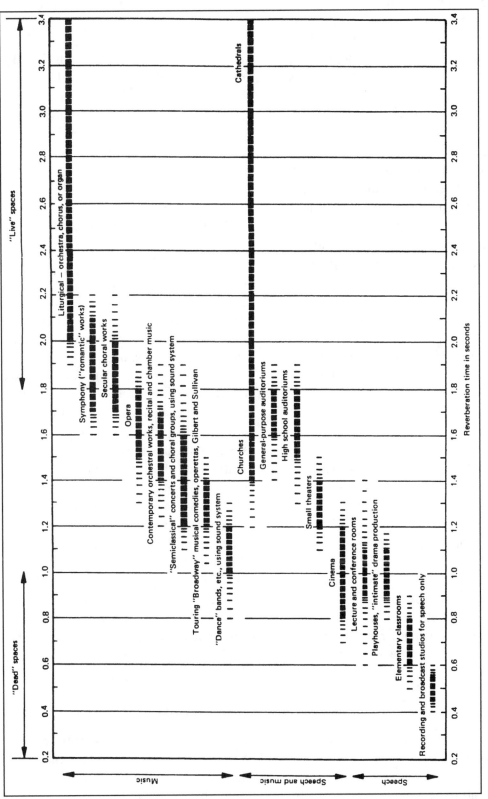

TABLE 8.5 – REVERBERATION TIMES

Reprinted courtesy of BBN Technologies.

As noted above, the reverberation time depends on the volume of the space and the absorptivity. Since the volume is generally determined by other factors, such as audience size, sight lines, building size, etc., the architect usually obtains the desired reverberation time by using materials that will provide the appropriate absorptivity.

Note that the audience itself represents a very large absorptive surface area, so a theater that is reverberant when empty becomes much less so when occupied. Using absorptive upholstery on the seats reduces the variation with audience size. Very reverberant spaces are often called "*live*" spaces, and spaces with short reverberation times are called "*dead*" spaces.

Example #4

What is the reverberation time of the room we calculated above, both before and after the addition of the drapes?

Solution:

$$T_R = .049\,V/A = .049 \times 20 \times 30 \times 10/283$$
$$= 1.04 \text{ seconds (before drapes)}$$

$$T_R = .049\,V/A = .049 \times 20 \times 30 \times 10/490$$
$$= 0.60 \text{ seconds (after drapes)}$$

If we were concerned about the reverberation time being a bit too long for speeches, we could provide acoustical tile on the ceiling, or add more drapes, or increase the absorptivity in some other way without changing the room size.

Room Acoustics

An auditorium or theater must be carefully designed to produce a satisfactory acoustical environment. The floor area is determined by the number of seats to be provided. As a general rule, good sight lines will also give good sound lines.

After selecting a reverberation time T_R to suit the purpose of the hall, the ceiling height of the hall must be established. A rule of thumb is to make the average ceiling height

$$H = 20 \times T_R$$

where
   H = height in feet
   T_R = desired reverberation time

The volume of the hall should be at least 100 cubic feet per person.

The two basic design goals are to reinforce reflections that arrive at the listener at nearly the same time as the sound from the source, and to cancel out reflections that would be excessively delayed. Thus the stage itself may have reflective surfaces, and the rear of the auditorium almost always has an absorptive surface or a shape that traps the sound. Sloping both the seating and the ceiling upward away from the stage is beneficial in two ways. The audience sight line is improved because the rear seat rows have visibility over those in front of them, and listeners in the rear can also receive sound by direct path from the stage.

Reflected sound from the ceiling is useful, but where speech intelligibility is important, the length of the reflected path should not exceed the length of the direct path by more than 34 feet.

Sound Transmission and Isolation

Besides reinforcing or controlling sound within a space, the most important consideration for the architect is limiting the communication of sound between two spaces. A certain level of

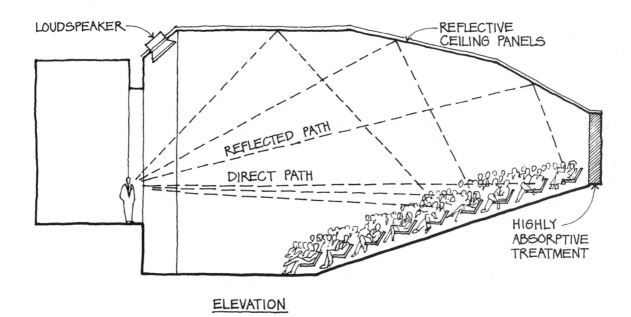

LOUDSPEAKER

REFLECTIVE
CEILING PANELS

REFLECTED PATH

DIRECT PATH

HIGHLY
ABSORPTIVE
TREATMENT

ELEVATION

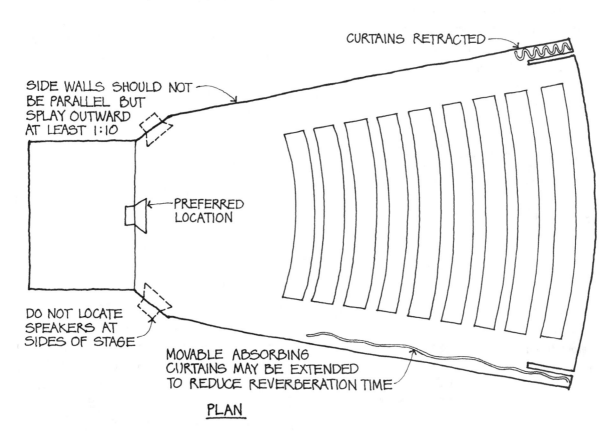

CURTAINS RETRACTED

SIDE WALLS SHOULD NOT
BE PARALLEL BUT
SPLAY OUTWARD
AT LEAST 1:10

PREFERRED
LOCATION

DO NOT LOCATE
SPEAKERS AT
SIDES OF STAGE

MOVABLE ABSORBING
CURTAINS MAY BE EXTENDED
TO REDUCE REVERBERATION TIME

PLAN

AUDITORIUM DESIGN FEATURES

quiet is required for the performance of many tasks, and excessive noise usually interferes with communication.

A family of curves, known as the 1957 Noise Criteria or NC curves, are widely used in specifying the maximum noise level in a given space under a given set of conditions. These are shown on the following page. If a given NC curve is specified, then the noise level in each octave band centered around the frequencies shown must not exceed the SPL level intercepted by the specified curve. For example, if the NC 30 curve is specified, the 250 Hz octave band must not exceed 41 dB as measured on a sound meter. To verify that the specified NC curve has been satisfied, octave band SPL readings must be made at each of the indicated frequencies and the results compared to the NC curve. If all of the measured values are equal to or below the intercepts of the curve, the curve has been satisfied. Table 8.6 indicates some typical NC requirements.

TABLE 8.6 – SUGGESTED NC CURVES		
Occupancy	**Minimum**	**Maximum**
Broadcast studio, concert hall	NC-15	NC-20
Bedrooms	NC-20	NC-30
Large conference room, courtroom	NC-20	NC-30
Small conference room, library, private office	NC-30	NC-35
General office area, stores, cafeterias	NC-35	NC-40
Minimum values provide sufficient background noise for masking, maximum values prevent outside interference.		

A similar set of curves, known as the 1971 Perceived Noise Criteria (PNC) has been proposed, which is used in much the same way as the NC curves.

Noise Reduction Through a Wall

An ideal wall separating two non-reverberant spaces would transmit some fraction (called τ) of the sound incident upon it (hitting it) in the source room to the other space. *Transmission loss (TL)* is defined as

$$TL = 10 \log 1/\tau$$

Thus, as τ increases, TL decreases.

In reverberant rooms the difference in IL between two real rooms separated by a real barrier is termed *Noise Reduction (NR)* and is related to the ideal TL by the equation:

$$NR = IL_1 - IL_2$$
$$= TL - 10 \log S/A_R$$

where
  TL = the free field transmission loss of the wall
  S = area of the separating wall, in square feet
  A_R = total absorptivity in the receiving room, in sabins

Example #5

Two identical 20 × 30 foot × 10 foot high rooms are adjacent to one another, and the IL in one room is 75 dB. What is the IL in the other room if they are separated by a shared 20 foot wall with a TL of 25 dB and the previously calculated absorptivity of the room is 490 sabins?

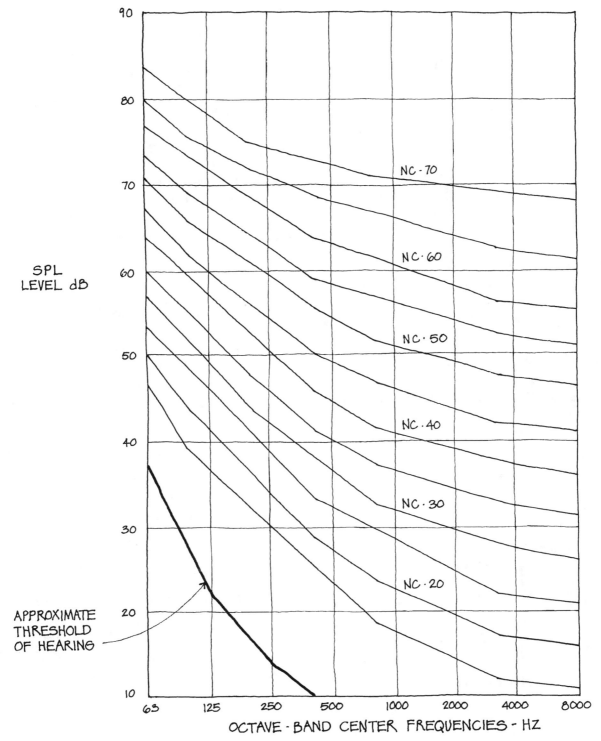

NOISE CRITERIA CURVES - 1957 NC

Solution:

$$NR = TL - 10 \log S/A_R$$
$$= 25 - 10 \log 20 \times 10/490$$
$$= 25 - 10 \log .408$$
$$= 25 - [10 \times (-.389)]$$
$$= 25 + 3.89$$
$$= 28.89 \ dB$$

$$IL_2 = IL_1 - NR$$
$$= 75 \ dB - 28.9 \ dB$$
$$= 46.1 \ dB$$

The TL of a wall is determined by its construction, its stiffness, and its mass. A general rule of thumb is that for every doubling of mass there is an increase of between 5 and 6 dB in TL.

Thus, one way to reduce the sound transmission of a wall is to increase its mass. For example, a concrete wall has much less sound transmission than a stud wall. However, concrete walls tend to be uneconomical.

Other ways of improving the TL of a wall include the use of staggered studs and flexibly mounting gypsum board, as shown below.

Sound always leaks through the weakest point in the isolating wall. Thus a concrete wall with an open window is like a bucket with a hole in it. Almost all of the sound goes through the window. It is of paramount importance, therefore, that holes for pipes, ducts, conduit, etc., be packed with insulation and sealed as well as possible.

It is important to understand the difference between transmission and reflection. Acoustical tile will stop reflection, but will not stop transmission. On the other hand, concrete stops transmission, but does not stop reflection. Do not confuse the processes. Acoustic tile and

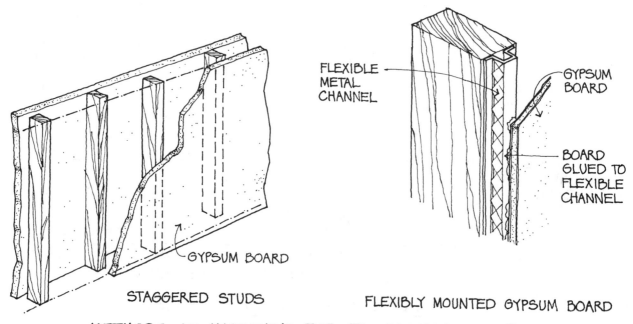

STAGGERED STUDS   FLEXIBLY MOUNTED GYPSUM BOARD

METHODS OF IMPROVING THE TL RATINGS OF WALLS

insulation may stop reflection *inside* a wall cavity, and thus reduce overall transmission. But applying acoustic tile to the outside surface of a wall does very little to reduce transmission. It reduces *reflection* back into the room, but does not stop *transmission*.

Sound Transmission Class

The *sound transmission class (STC)* is a widely accepted method of rating walls, doors, etc. in terms of their typical or overall resistance to sound transmission. Obviously, the transmission of different sound frequencies varies from wall to wall, so a weighted average of all frequencies is used.

The STC rating of a given wall section is established by measuring the TL of a test panel at 16 one-third octave bands and plotting these as in the following drawing. The standard STC contour is fitted, as closely as possible, to the curve. Then the 500 Hz dB value of the standard curve is used as the STC rating of the test panel.

The shape of the standard STC contour was chosen to be representative of a nine inch thick brick wall. Stud walls and other nonconforming walls do not fit as well, and are not as accurately represented by the measurement.

STC values are widely quoted in catalogs and advertising, and are useful for design. The architect should use these values with some caution because actual construction practices tend to be less perfect than those used to construct the test panel.

Impact Noise

Impact noises are erratic sounds caused by footfalls, dropped objects, the vibration of mechanical equipment, etc. The resulting vibration of the structure is then manifested as airborne sound radiated from other locations.

A standardized method of measuring the degree of isolation of impact noise in the structure

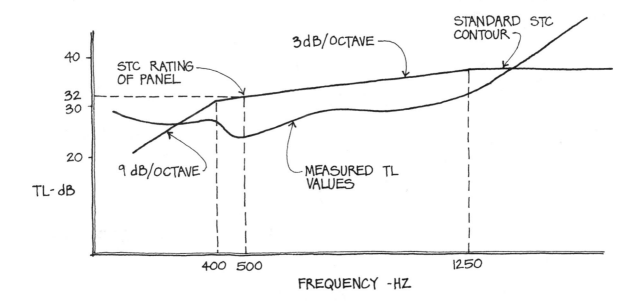

STC DETERMINATION OF A PANEL

has been developed. The method utilizes a special "tapping machine" that is placed on the floor to be tested. A sound meter with one-third octave filters is located in the room beneath the tapping machine and is used to measure SPL at each frequency. The resulting data is then plotted, and an *impact isolation class (IIC)* contour is fitted in accordance with prescribed rules. The 500 Hz intercept of the IIC contour becomes the IIC rating of the floor/ceiling assembly. The IIC ratings are utilized similarly to STC values.

IIC values can be improved in several ways. Carpeting and resilient tile floor coverings are effective at middle and high frequencies. Suspended ceilings are also very effective at middle and high frequencies, especially if combined with carpeting. A concrete slab floated on compressed glass fiberboards laid on the structural floor is very effective in improving the performance at all frequencies and, thus, the IIC rating.

Code Requirements

In addition to OSHA requirements for noise levels at job sites, the Uniform Building Code has established minimum airborne and impact sound isolation values for residential occupancies. Dwelling units and guest rooms must have some acoustical privacy between each other and between them and public areas, such as corridors and service areas. General guidelines are STC 50 for walls, floors, and ceilings, and STC 26 for entrance doors.

Speech Privacy

Speech privacy is an important factor of concern in offices, apartment houses, etc. Conversations overheard from the other side of a partition are annoying, and also cause concern that one's own conversations may be overheard beyond the walls. One of the factors that determines whether overheard speech is intelligible is the background level of noise. This gives the designer an additional tool to employ in creating privacy: increasing the background noise level in an innocuous way. This is often called masking noise or white noise. The rushing sound of air from HVAC outlets can provide such sound, because it represents a good mix of

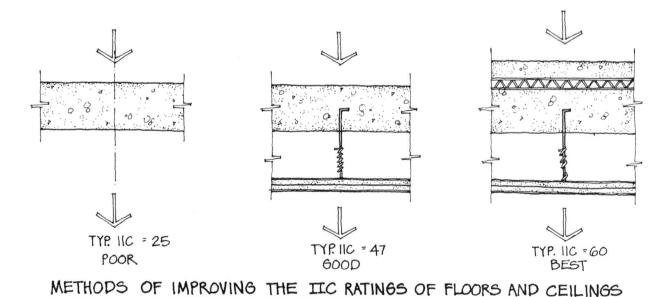

TYP. IIC = 25
POOR

TYP. IIC = 47
GOOD

TYP. IIC = 60
BEST

METHODS OF IMPROVING THE IIC RATINGS OF FLOORS AND CEILINGS

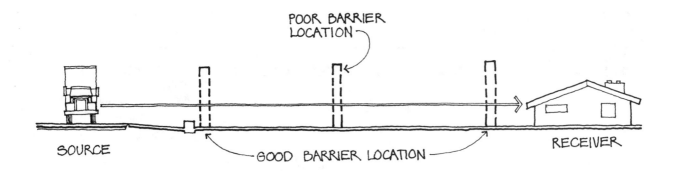

OUTDOOR SOUND BARRIER

all frequencies. Do not confuse this with the knocks, squeaks, and vibrations sometimes present in mechanical equipment, which tends to mask little, and is a source of annoyance.

SPECIAL CASES

Outdoor Sound Barriers

There is frequently a need to control outdoor sounds, as from a highway near a structure. Solid barriers, as shown above, can be effective, but they must be properly located. The best location for a barrier is very close to the source or very close to the receiver; middle positions are significantly less effective. The barrier must be higher than the line-of-sight between the source and the receiver; the higher the better. Unfortunately, the attenuation of a barrier is not constant, but instead increases at 3 dB per octave. For this reason, low-pitched noises, such as truck engine sounds, are not reduced as much as tire whine, which is at a higher frequency.

Attenuation values of 10 dB or more at 500 Hz are possible; the upper limit is about 25 dB at the higher frequencies.

The use of trees and vegetation as an acoustical barrier was once thought to be effective.

However, experiments show that they provide almost no attenuation at all for transmitted sounds. Vegetation in front of a barrier may reduce the reflected sound, but vegetation alone does very little to stop the transmission of sound.

Mechanical Systems

Control of noise generated by mechanical systems is an important aspect of modern design. Most mechanical equipment produces noise having a definite frequency or pitch related to the rotational speed of the equipment. Of course, numerous harmonics also exist in most cases, producing a larger band of frequencies from any given source. Typical sources are motors, fans, pumps, and compressors. Moving fluids, such as air and water streams, also create noise due to turbulence in the flow, and the noise generated by such flow has no definite frequency; instead, it contains a random mixture of essentially all frequencies. This is what we referred to previously as *"white noise."*

A low background level of white noise can mask other sounds, particularly in the frequency range of speech. This provides privacy by making neighboring conversations unintelligible when their volume drops below that of the white noise.

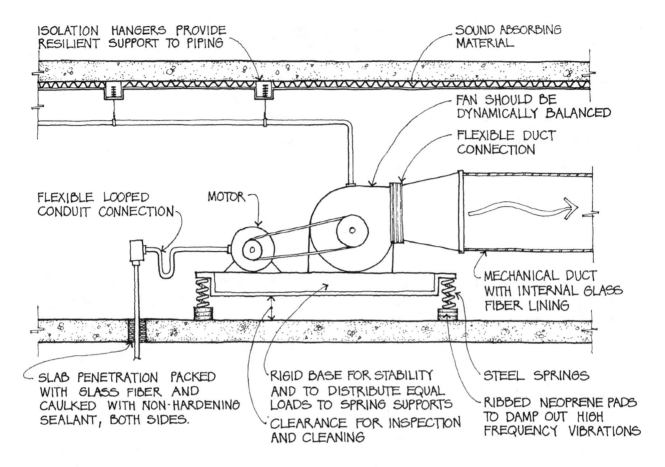

ISOLATION HANGERS PROVIDE RESILIENT SUPPORT TO PIPING

SOUND ABSORBING MATERIAL

FAN SHOULD BE DYNAMICALLY BALANCED

FLEXIBLE DUCT CONNECTION

FLEXIBLE LOOPED CONDUIT CONNECTION

MOTOR

MECHANICAL DUCT WITH INTERNAL GLASS FIBER LINING

SLAB PENETRATION PACKED WITH GLASS FIBER AND CAULKED WITH NON-HARDENING SEALANT, BOTH SIDES.

RIGID BASE FOR STABILITY AND TO DISTRIBUTE EQUAL LOADS TO SPRING SUPPORTS

CLEARANCE FOR INSPECTION AND CLEANING

STEEL SPRINGS

RIBBED NEOPRENE PADS TO DAMP OUT HIGH FREQUENCY VIBRATIONS

DESIGN FEATURES TO CONTROL MECHANICAL EQUIPMENT NOISE

The use of high quality mechanical components will help to reduce system noise, but the designer must still pay attention to details of design to insure satisfactory sound levels in the completed structure. The motor and its load must be rigidly mounted on a common sub-base to insure shaft and/or belt alignment, and the sub-base should be isolated from the structure by springs and pads of rubber, cork, or neoprene. The resonant frequency of the sub-base, motor and machine assembly should be one third, or less, of the frequency of the motor. Concrete is sometimes added to increase mass and lower the resonant frequency. Of course the conduit, pipes and duct work must be equipped with flexible

connections to avoid transmitting mechanical vibrations into the remainder of the system. Some of those features are shown in the illustration above.

Long ducts may be acoustically lined to control sound waves through the ducts, but where sound must be controlled in short duct sections, mufflers as short as three feet are available with good attenuation over a wide range of frequencies. Other sound control features include shock arrestors on water pipes to control water hammer and resilient packing to seal around pipes and ducts where they pass through structural openings.

SUMMARY

We have discussed the behavior of sound, which is similar to light, in that it follows the inverse square law, and may be transmitted, absorbed, or reflected. We have also shown examples of calculations using the logarithmic relationships typical of acoustics. We have discussed the sensitivity of the human ear, and code and OSHA regulations designed to protect the ear. We have covered the behavior of sound within an enclosed space and through walls between spaces.

The coverage of acoustics in this lesson should provide the architectural candidate with the basic understanding that he or she is expected to have.

LESSON 8 QUIZ

1. What criteria are used in specifying the maximum noise level in a given space under a given set of conditions?

   A. PWL   C. NC

   B. NRC   D. NR

2. What is the logarithmic measure of the intensity of a sound relative to a reference intensity in watts/m^2 or watts/cm^2?

   A. IL   C. STC

   B. SPL   D. NC

3. Which of the following is a method of rating wall sections or doors according to their typical or overall resistance to sound transmission?

   A. SPL   C. PWL

   B. STC   D. NR

4. Reverberation time is

   I. the time required for a sound to decay 60 dB in a space.

   II. the time it takes for an echo to return.

   III. longer in a dead space, and shorter in a live space.

   A. I only   C. I and III

   B. II and III   D. I, II, and III

5. The maximum acceptable IL for a hospital is

   A. 100 dB.   C. 40 dB.

   B. 70 dB.   D. 10 dB.

6. Doubling the distance from the source to the receiver would result in a

   A. drop of 6 dB.

   B. gain of 6 dB.

   C. drop of 3 dB.

   D. gain of 3 dB.

7. Doubling the number of sources at a given intensity would result in a

   A. drop of 6 dB.

   B. gain of 6 dB.

   C. drop of 3 dB.

   D. gain of 3 dB.

8. Doubling the mass of a theoretical wall causes its sound transmission to

   A. decrease 6 dB.

   B. increase 6 dB.

   C. decrease 3 dB.

   D. increase 3 dB.

9. Given a concrete room with concrete walls, floors, and ceiling, adding carpeting and drapes does which of the following?

   I. Increases the absorptivity of the room

   II. Decreases the reverberation time of the room

   III. Decreases the intensity of the sound that a small motor would make in the room

   A. I only   C. I and III

   B. II and III   D. I, II, and III

10. A partition with a sound transmission class rating of 55 dB would be

   A. very good at stopping reflected sound.

   B. barely acceptable at stopping transmitted sound.

   C. very good at stopping transmitted sound.

   D. barely acceptable at stopping reflected sound.

VERTICAL TRANSPORTATION

INTRODUCTION

The first architect who decided to build up, rather than out, might have made that decision more than 5,000 years ago. Regardless of the precise moment in history, from the time that multistoried structures first appeared on the architectural scene, builders have had to consider how to get people from one level to another.

The earliest forms of vertical transportation were fairly primitive ramps, ladders, and stairs.

In those isolated cases that involved mechanical ingenuity, the device was generally intended to move materials, rather than people. As an example, the ancient Romans used a movable platform hoist that operated in a shaftway containing guides. Some of these hoists were powered by hand-driven windlasses, and others used animal power; but in either case, the hoist was used to lift materials during construction or to transport products to elevated stories or even over steep mountain cliffs.

During the early 19th century, steam-driven hoists replaced human and animal labor, but they did little more than their ancient counterparts. These early hoists were rarely used to transport people, principally because they lacked any kind of safety device, and as a result, they were extremely hazardous. Hoisting ropes were made of fiber, and if the rope failed, as it often did, there was no way to stop the hoist from crashing to the ground.

In 1853, Elisha Graves Otis developed an elevator safety device that ultimately altered the shape of cities throughout the world. The device consisted of a pair of spring-loaded elements that meshed with ratchets in the guide rails when the tension of the hoisting rope was released. Otis demonstrated his safety device by ascending in a hoist that was part of an exhibit at the New York Exposition of 1853. Halfway up in the open-sided shaft, an assistant suddenly and dramatically sliced the hoisting rope with an axe, but the platform held steady. The gathered crowd stood astonished as it witnessed the birth of the safety elevator. By the early 1900s, elevators served practically every new multistoried commercial structure, and in subsequent years, improvements in elevatoring made possible the 100-story structures of today.

Other forms of vertical transportation widely used in contemporary construction are the moving inclined ramp and the moving stairway. The moving stairway was developed by Otis' company and first introduced to the world at the Paris Exhibition of 1900. Although it caused quite a sensation, it was a far cry from the sophisticated escalators that are found in most of today's department stores.

Vertical transportation refers to all the systems used to move people and materials vertically. These can be divided into two categories: those that require human effort, such as stairs, ramps, and ladders; and those that are mechanically powered, such as elevators, escalators, moving ramps, dumbwaiters, and vertical conveyors. This lesson begins with stairs, ramps, and ladders, and then examines mechanically powered devices.

STAIRS

Stairs are the most common means of vertical circulation between floors of a building, and their design is dictated by considerations of safety and convenience. Stairways may be as narrow as 36 inches, but when serving an occupant load of 50 or more, they must be no less than 44 inches wide. The rise of a step should not exceed 7 inches, and the run of a tread should not be less than 11 inches. In addition, safe practice dictates maintaining one of the

following conventional relationships between riser and tread:

Riser + Tread	= 17 to 17.5 inches
Riser × Tread	= 70 to 75 inches
2 Risers + Tread	= 24 to 25 inches

Stair landings should be as long as the stairway width (up to 44 inches maximum), the vertical distance between landings may not exceed 12 feet, and headroom clearance should never be less than 6'-8".

Fire stairs are required for emergency exiting, and their number is controlled by the occupancy of the building and the occupant load of each floor, as specified in the building code. In general, all such stairways must be enclosed with fire-resistive walls and be equipped with complete fire assemblies, including fire doors that open in the direction of emergency travel.

Monumental stairs, often referred to as decorative stairways, are often used in larger buildings to connect the main floor to a mezzanine or

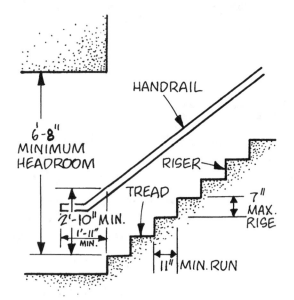

CRITICAL STAIR DIMENSIONS

second floor. Stairways more than 88 inches in width require intermediate handrails, but in any event, monumental stairs are rarely permitted to be used as legal fire stairs.

Residential stairs are governed by more lenient regulations than are fire stairs in public buildings. Because of their considerably smaller occupant load, residential stairs may be narrower, only one handrail is required, fire-resistive construction is not mandatory, and winders and spiral stairs are sometimes acceptable.

RAMPS

Ramps are inclined walkways that allow easy vertical transition between different levels. The slope of ramps may be summarized as follows:

1:20 (5.0 percent) or less, not considered to be a ramp

1:15 (6.7 percent) or more, requires handrails

1:12 (8.3 percent) maximum slope permitted for access for persons with disabilities

1:8 (12.5 percent) maximum slope permitted by building code (typically in theater aisles)

Ramps are usually as wide as the corridors leading to them, except for handicapped ramps, which must be a minimum of 36 inches wide.

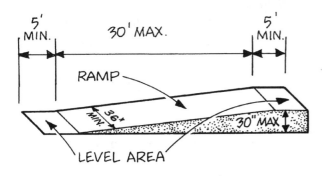

HANDICAPPED RAMP DIMENSIONS

In addition, handicapped ramps have a maximum rise of 30 inches, and landings must be at least five feet long. Handrails are required on handicapped ramps if the ramp is more than six feet long or has a rise greater than six inches.

LADDERS

Ladders are generally used for access to roofs or in utility and service areas where space is tight and traffic is minimal. Although occasionally made of wood, most permanent ladders in buildings are fabricated from metal. Vertical ladders should be a minimum of 18 inches wide, have rungs spaced 12 inches apart, and be installed at least 6 inches from walls to allow for adequate toe space.

ELEVATORS

Elevators are movable enclosures that provide vertical transportation for people and freight. They generally consist of an enclosed platform, or car, which is raised or lowered along rails within a vertical shaft, and the mechanical equipment that provides power, control, and safety for the operation.

The two major types of elevators in common use are *hydraulic* and *electric*. Each type has distinctly different characteristics and uses. Hydraulic elevators are directly ram-driven from below, while electric elevators employ steel cables with counterweights and operate on the traction principle. In other words, the hydraulic elevator platform is *pushed* up by a rod, whereas the electric elevator platform is *pulled* up by cables.

Hydraulic Elevators

Hydraulic elevators are commonly used for low-rise buildings (about 50 feet or five stories

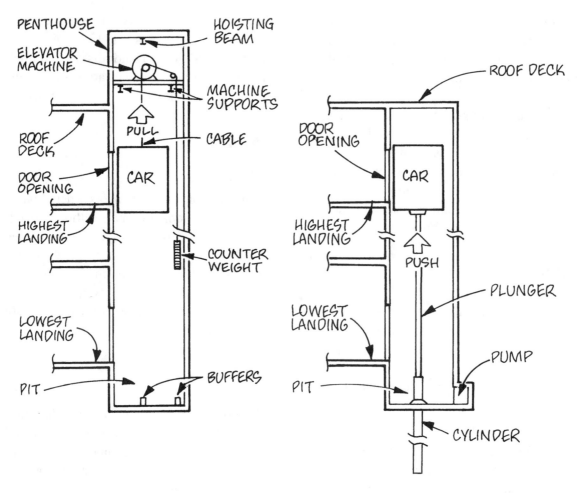

ELECTRIC ELEVATOR

HYDRAULIC ELEVATOR

maximum), and have speeds that vary between 25 and 150 feet per minute (fpm). Because of their short travel distance at relatively low speeds, hydraulic elevators are commonly used for freight in industrial and low-rise commercial buildings; for passengers in garden apartments and motels; and occasionally in single-family residences.

Hydraulic elevator platforms sit over a plunger, or ram, which operates in a cylinder that usually extends as far into the ground as the elevator rises. Oil serves as the pressure fluid, the supply of which is controlled by high-speed pumps.

Because of their simplicity of operation, hydraulic elevators cost less than electric elevators. In addition, they have no wire cables or overhead machinery, and consequently, they do not require a penthouse. Hydraulic elevators are especially suited to carry freight; single ram capacities are available up to 10 tons, and multiple rams have been used to carry up to 50 tons.

Electric Elevators

Electric elevators are generally used in commercial and institutional buildings greater than 50 feet in height. They have capacities up to 10,000 pounds and may attain speeds, in buildings over 50 stories, of up to 1,800 fpm.

Traction is used to transmit lifting power to the hoisting cables of an electric elevator. This is achieved by means of the friction that develops when the cables run over grooves in the machine-driven sheave. On one end of the cables is the elevator car, and on the other end are counter-weights, weighing about 40 percent of the load capacity, which are used to reduce power requirements.

The motor and drum assembly that moves the elevator car is called a traction machine, and it may be gearless or geared. The gearless machine has the motor, sheave, and brake all mounted on a common shaft; thus one revolution of the motor turns the main sheave one revolution. Geared machines, on the other hand, have the motor and brake on one shaft, which in turn drives a second main shaft. The two shafts are connected by means of gears, so that the motor shaft may make several revolutions for each turn of the main shaft. Gearless machines are used for high-speed installations, while geared machines, with their greater flexibility, are preferred for low-speed applications.

Roping

Roping for elevators has a considerable effect on the loading and stress of the cables, machine bearings, and building members. Traction-type machines are classified as either single-wrap or double-wrap. In the single-wrap design, the cables pass only once over the elevator machine sheave. For high-speed service, additional traction may be required, and this is obtained by using the double-wrap design. In this case, cable life is usually shorter, since there are more bends in the cable. Elevators are roped either 1:1 or 2:1, referring to the distance the cable moves in relation to the distance the car moves. In 2:1 roping, the car speed is one-half that of the cable; however, 2:1 roping requires that only half the weight be lifted by the motor, therefore

allowing a smaller elevator motor to be used. Illustrated on the following page are several elevator roping arrangements.

Safety Features

Safety features on elevators are numerous and highly refined, since the potential hazards associated with elevator use can be critical. The main brake is mounted on the motor shaft of the elevator machine and operated automatically by the control panel. The brake is self-applying, so that the car will be stopped in the event of a power failure.

The *governor* measures and limits the elevator speed by means of the control panel. In case of excess speed, it actuates the safety rail clamp. Top and bottom limit switches will also actuate the safety rail clamp if the elevator attempts to exceed its proper range of travel. The safety rail clamp is, in effect, another type of brake. It consists of metal jaws on the car that, in case of emergency, clamp the guide rails and hold the car in a fixed position. If a car is disabled between floors, an emergency exit hatch in the roof of the car permits access or egress.

Car bumpers are located at the bottom of the shaft to stop the car if it should overtravel at low speed. Their purpose is to arrest the car's motion more gently than a solid stop; however, they are not designed to stop a free-falling car.

The doors of an elevator car have safety edges consisting of a spring-loaded lip. If this lip encounters a person, package, and so forth, the door will reopen and prevent the elevator from moving. In addition, automatic (unattended) elevators have an electric eye system that prevents the door from pinching passengers or objects. Car and door interlocks are provided at each floor to prevent the possibility of a lobby door opening when the elevator is elsewhere. Cars are also

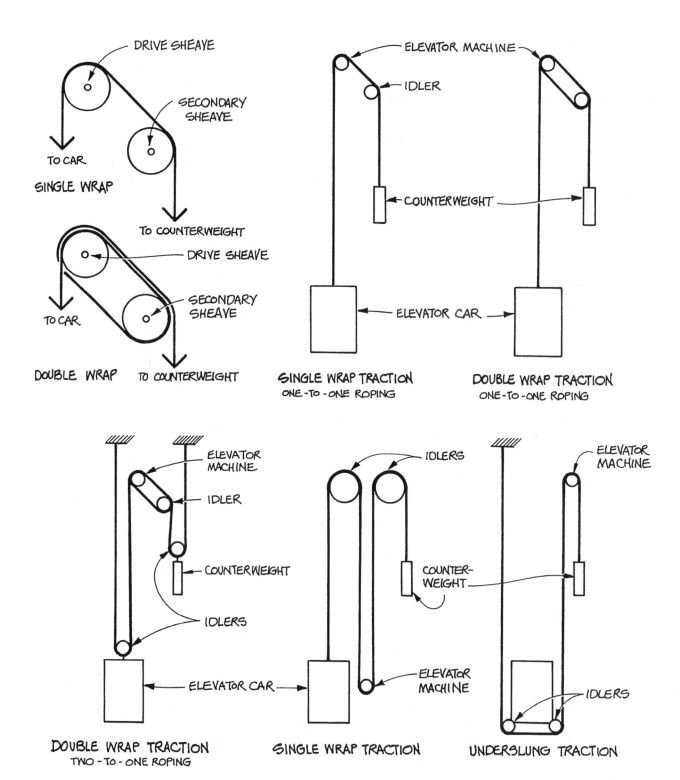

DRIVE SHEAVE

SECONDARY SHEAVE

TO CAR

SINGLE WRAP

TO COUNTERWEIGHT

DRIVE SHEAVE

SECONDARY SHEAVE

TO CAR

DOUBLE WRAP   TO COUNTERWEIGHT

ELEVATOR MACHINE

IDLER

COUNTERWEIGHT

ELEVATOR CAR

SINGLE WRAP TRACTION
ONE-TO-ONE ROPING

DOUBLE WRAP TRACTION
ONE-TO-ONE ROPING

ELEVATOR MACHINE

IDLER

COUNTERWEIGHT

IDLERS

DOUBLE WRAP TRACTION
TWO-TO-ONE ROPING

ELEVATOR CAR

IDLERS

COUNTER-WEIGHT

ELEVATOR MACHINE

SINGLE WRAP TRACTION

ELEVATOR MACHINE

IDLERS

UNDERSLUNG TRACTION

ELEVATOR ROPING ARRANGEMENTS

equipped with automatic leveling devices to reduce the hazard of passengers tripping at lobby thresholds.

Elevators are not considered legal exits. In case of fire, automatic controls generally close the elevator doors and return the cars to the lower terminal, where the doors open and the elevators become manually operable only by a key.

Capacity and Speed

The capacity and speed of elevators vary considerably, depending on their use and the height and overall quality of the project. Normally, a small office building requires a minimum-size elevator that is rated at 2,500 pounds capacity and has a car 5 by 7 feet in size. This allows the use of a 3'-6" wide center-opening door. Passenger elevators, even in heavily-traveled prestige office buildings or monumental institutional buildings, rarely exceed 4,000 pounds capacity.

Elevator speeds should be adequate to provide prompt, efficient service. The maximum usable speed is limited by the building height, since acceleration and deceleration require a substantial distance. In other words, elevator cars, without using "jack-rabbit" starts or abrupt stops, are unable to attain or reduce a high speed in a short travel span. As an approximate rule of thumb, the rated elevator speed may be figured as 1.6 times the rise in feet, plus 350. Thus, with a 300-foot-high building, a speed of about 800 fpm would be called for $(1.6 \times 300 + 350)$. Listed below are recommended elevator speeds for office or general purpose buildings.

Operating Systems

Operating systems used in elevators today are almost all automatic types, replacing the older, operator-controlled car switch type. In the single automatic operation, the car responds to the first button pressed, ignoring all other calls until the car reaches its first destination. Elevators using this operating system are unable to store calls, but their use in apartments and hospitals is quite acceptable.

With the collective operating system, calls may be stored. Selective-collective systems answer all calls in the direction of a car's travel; that is, when traveling up, the car will not stop at a floor where a down button has been pressed. Fully automatic systems are generally used for tall office buildings. They have all the advantages of the selective-collective system, but in addition, service can be adjusted for varying traffic conditions. For example, during early morning hours, cars can be regulated to operate at fixed

| | Recommended Speeds in Feet per Minute | | | |
Height in Floors	Small	Average	Prestige	Service
2–5	200–250	300–350	350-400	200
5–10	300–350	350–500	500	300
10–15	500	500–700	700	350–500
15–25	700	800	800	500
25–35	—	1000	1,000	500
35–45	—	1,000–1,200	1,200	700–800
45–60	—	1,200–1,400	1,400–1,600	800–1,000
over 60	—	—	1,800	1,000

intervals, or when loaded; and after reaching the highest floor, the car returns to the entrance lobby. In tall buildings, elevators may be zoned, with each car serving only a specified group of floors. This permits high-speed operation for a portion of the trip and generally results in more efficient use of elevator equipment.

Architectural Considerations

Architectural considerations in the use of elevators include the determination of elevator size, number, and location for efficient traffic. These requirements depend on building population and function, building height and number of floors, volume of traffic, car capacity, and speed of elevator operation.

Elevator traffic is generally computed for peak or critical traffic periods, which vary with the type of building. Office buildings might reach their peak period during lunch time, whereas the peak period in hospitals might occur during visiting hours. Traffic is measured by the number of persons handled during a five-minute peak interval. When this figure is divided by the five-minute handling capacity of an elevator, the minimum number of elevators can be determined.

The handling capacity of an elevator is computed on the basis of both its car size (the number of people carried) and its round-trip time. Normally, an average person requires two square feet of space in order to feel comfortable in an elevator car. During rush hours, however, this may be reduced to as little as 1.3 square feet.

The round-trip time comprises all of the factors involved with a full-speed, non-stop, round-trip run as shown in the example in the right column.

In general, time intervals between elevators should be between 20 and 30 seconds; a waiting time of more than 35 seconds is usually considered unsatisfactory.

Action During Round-Trip Run	Time in Seconds
1. Passenger operates call button, doors open	3
2. Passenger enters car, operates floor button	2
3. Doors close	3
4. Car travels to upper floor	7
5. Doors reopen (starting before car stops) 3 less 1 premature opening =	2
6. Passenger leaves car	2
Total time to serve one passenger	19
7. Doors close	3
8. Car returns to original floor	7
Total Round-Trip Time	29

The *location of elevators* is critical to the efficiency of pedestrian traffic flow in a building. Elevators should be centrally located and easily accessible from all building entrances. The best arrangement is to place the elevators in an alcove off the main corridor, to reduce traffic conflicts. With groups of up to three elevators, the cars are best placed in a row; with four elevators or more, they are best grouped in banks opposite one another. In general, when more than eight elevators are needed, the best arrangement is a combination of local and express traffic banks.

Door arrangements on elevator cars may vary somewhat, with adequate width and rapid operation being necessary for efficiency. Very small apartment house elevators may have swinging doors, but higher-capacity installations require sliding doors to facilitate passenger loading. Doors that open to one side may be either single-speed or two-speed, the latter having a second panel telescoping alongside the first.

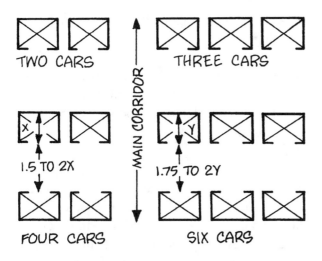

ELEVATOR CAR ARRANGEMENTS

Center-opening doors may also be single-speed or two-speed, with the latter arrangement comprising a total of four panels. Center-opening doors operate more efficiently than side-opening doors, since they make possible speedier passenger loading and unloading. In general, the optimum-sized door is 3'-6" wide, which allows two persons to enter or leave an elevator simultaneously.

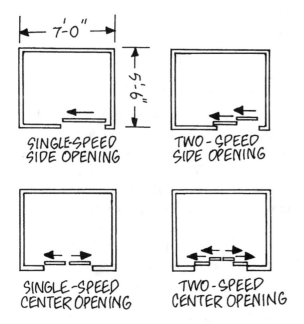

ELEVATOR DOOR ARRANGEMENTS

Special design considerations may be required for elevators, depending on the unique function of a building. For example, hospital elevator cars should be large enough to accommodate beds, wheelchairs, and stretchers (5'-4" × 8'-4"), and have a capacity of 3,500 pounds or more. Restaurants and department stores require separate passenger and service elevators; and for industrial buildings, shift changes and the movement of large equipment must be taken into account. Other special situations include hotels (baggage); prisons (security); theaters (peak loads); professional buildings (high visitor population); and courthouses (user separation).

FREIGHT ELEVATORS

Freight elevators are used to vertically transport equipment, materials, and goods, rather than passengers. In low-rise buildings, hydraulic elevators are appropriate, but electric elevators are generally more economical for lifts exceeding 50 feet. Generally, freight elevators have a less finished appearance, lower speed, and greater capacity than passenger elevators. Depending on the building height, standard speeds vary between 75 and 200 fpm, but they may be greater for smaller capacity elevators. Freight elevators are available in capacities from 2,500 to 25,000 pounds or more; some can carry fully loaded trailer trucks and even railroad cars.

The American Standard Safety Code for Elevators defines the classes of freight elevators as follows:

Class A General Freight. Loaded by hand truck. No single piece of freight to exceed one-quarter capacity load of elevator, based on 50 pounds per square foot of platform area.

Class B Motor Vehicle Garage Elevator. Automobiles and trucks only, based on capacity of 30 pounds per square foot of platform area.

Class C Industrial Truck Loading. Must be able to support impact load and weight of truck. Capacity not less than 50 pounds per square foot of platform area.

Service elevators used in residential, commercial, and institutional buildings are usually passenger elevators that have been modified to handle oversize loads or hand trucks. They commonly have abuse-resistant interiors and are equipped with horizontal sliding doors, in contrast to the vertical biparting doors used in freight elevators. Service elevators must comply with code requirements for carrying passengers, and their size is generally determined by the largest or bulkiest load that is anticipated to be carried.

Sidewalk elevators are freight or supply lifts that rise to an upper level by opening hatch doors located in the ground floor. Years ago these were commonly placed in sidewalk areas; however, most municipal codes prohibit this practice today. Sidewalk elevators are either hydraulic or electric, two stories maximum, and they travel at about 25 fpm.

DUMBWAITERS

Dumbwaiters are hoisting and lowering devices, very much like small elevators, used to vertically transport materials and supplies, but never people. They are distinguished from freight elevators by limitations found in elevator safety codes; namely, car platforms may not exceed nine square feet in size, nor can the car height exceed four feet. Thus, dumbwaiters are quite small: their capacity never exceeds 500 pounds. They may be automatically powered (controlled by push buttons from the landing, not the cab), or they may be manually operated by pulling on ropes. In either case, they have very few safety features, as their principal purpose is material handling.

Dumbwaiters are used for a variety of service-related functions, such as distributing mail in office buildings, delivering books in libraries, and transporting food and supplies in restaurants and hospitals.

Vertical conveyors are used to distribute a continuous flow of materials (never people) throughout multistoried buildings. In several different ways they bear a resemblance to other forms of vertical transportation. For example, they are mechanically similar to escalators in that they operate with a continuous chain which is driven by an electric motor. They are also similar to dumbwaiters in that they are small, carry relatively light loads, and operate at slow speeds. Vertical conveyors, like elevators, are enclosed in fire-resistant shafts.

Vertical conveyors carry trays or shelves spaced at regular intervals along the conveyor, with up loads on one side and down loads on the other. In most cases they are used to distribute paper work or light supplies.

ESCALATORS

An escalator, sometimes called a moving stairway, is a mechanical device on which passengers are transported along an incline from one floor to another. They provide rapid, comfortable, and continuous vertical travel without the long waiting periods frequently associated with elevators. Escalators are used to move large numbers of people from floor to floor quickly, efficiently, safely (safer than stairs), and at a relatively low cost of operation.

Escalators are made up of custom-built steel trusses that are fitted between floors, an

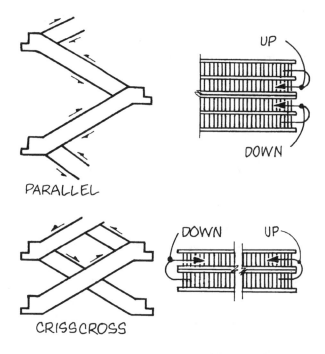

PARALLEL

CRISSCROSS

ESCALATOR ARRANGEMENTS

endless belt that contains the steps, and synchronized handrails. The normal angle of incline is approximately 30 degrees, and the operating speed is either 90 fpm or 120 fpm along the incline.

Escalators are rated by speed of operation in feet per minute and by nominal width, which is measured at approximately hip level. The two standard widths are 32 and 48 inches; the first allows a parent and child to share the same step, while the second allows two adults to ride side-by-side. Escalator capacity, expressed in passengers per hour, ranges from 4,000 to 5,000 passengers per hour on a 32-inch-wide 90 fpm escalator, to 8,000 to 10,000 passengers per hour on a 48-inch-wide escalator traveling at 120 fpm.

Because escalators are continuously running and unidirectional, pairs of escalators are necessary for two-way service, one for up traffic

and one for down traffic. The conventional arrangements used are either parallel or criss-cross; the latter is preferred, as it is a more compact arrangement with lower structural requirements and hence lower cost.

Escalators are not generally accepted as legal exits, as their unidirectional travel may increase, rather than reduce, the hazards of emergency exiting. Furthermore, escalators are difficult to enclose, although they may be made fire-resistant with rolling shutters and deluge-type sprays.

As with elevators, escalators should be located so that most people entering a building can see them. They should be in the path of the heaviest expected traffic, centrally located, and with ample space at floor levels. For years, department stores have utilized escalators to move large groups of shoppers and steer them towards specific merchandise. However, in recent times escalators have also been installed in transportation terminals, convention halls, hotels, office buildings, theaters, sports arenas, and other buildings that must transport large numbers of people in a minimum amount of time.

Moving ramps are a form of vertical transportation closely related to escalators; the principal difference, of course, is that moving ramps have a continuous tread, rather than individual steps. They incline up to about 15 degrees, and when flat, they are referred to as moving sidewalks. Moving ramps are about 40 inches wide and have speeds that vary between 140 and 180 feet per minute. They allow pedestrians to move swiftly, as one may continue to walk along the moving treadway, doubling one's speed. Moving ramps provide an effective means of transporting and directing large numbers of people quickly, as for example, at airport terminals; and their use appears to be increasing.

CONCLUSION

The future of elevatoring promises to be imaginative and exciting, if the recent past is any indication. Over the last several years, a new concept of elevatoring multipurpose buildings has developed, in which separate groups of elevators serve the different areas of a building. For example, in the John Hancock Center in Chicago, one group of elevators serves the 40 stories of offices below, while a separate group is used for the 44 stories of apartments above.

A method of reducing space requirements in tall buildings is the use of double-deck elevators traveling in a single vertical shaft. The two compartments of the elevator are loaded simultaneously; for example, during early morning rush hours, passengers destined for odd-numbered floors enter the bottom deck, and those for even-numbered floors enter the upper deck. When the elevator stops, passengers from both decks are discharged simultaneously. Elevator controls have also moved into the computer age, with microcomputers that continuously scan the system to make it more efficient.

Dramatic and decorative glassed-in elevator cars are also currently used in a number of installations where the view in motion provides a thrilling experience. In contemporary hotels, such as the Hyatt Regency Hotels in Atlanta and San Francisco, occupants of large atrium-lobbies share the excitement of movement, as spectators, with the elevator passengers. Similar observation-type elevator cars are sometimes used on the exterior of buildings, providing the passengers with panoramic views of the urban landscape.

LESSON 9 QUIZ

1. The maximum usable speed of an elevator is generally limited by
   A. considerations of safety.
   B. car capacity.
   C. counterweight/load ratio.
   D. building height.

2. In the design of a three-story commercial building with a maximum height of 45 feet, the architect wishes to have a constant roof silhouette with no projections above the 4-foot-high parapet. Consequently, the architect should choose
   A. an electric elevator with geared traction and operating machinery at the basement level.
   B. a gearless electric elevator with operating machinery in a four-foot-high roof structure.
   C. a hydraulic elevator with operating machinery in a four-foot-high penthouse.
   D. a hydraulic elevator with operating machinery located in the elevator pit.

3. The recommended speed of a passenger elevator in a 20-story building should be about
   A. 200 feet per minute.
   B. 300 feet per minute.
   C. 800 feet per minute.
   D. 1,000 feet per minute.

4. Every fire door that is part of a fire stairway exiting system must open
   A. in the direction of travel.
   B. into the enclosed stairway.
   C. toward the flight leading down.
   D. toward the building exterior.

5. In which of the following ways do freight elevators generally differ from passenger elevators?
   I.   They are never used by people.
   II. They are slower.
   III. They have greater capacity.
   IV. They require fewer safety features.
   V.   Their doors operate differently.
   A. I, II, and III   C. II, IV, and V
   B. II, III, and V   D. I, III, and IV

6. In order to connect two levels whose difference of elevation is two feet, a handicapped ramp must be at least
   A. 10 feet long.   C. 24 feet long.
   B. 12 feet long.   D. 30 feet long.

7. What safety device on an electric elevator provides life safety protection in the event of a power failure?
   A. A governor
   B. An automatic brake
   C. The interlocks
   D. The car bumpers

8. The principal advantage in the use of moving ramps is

   A. speed.
   B. comfort.
   C. safety.
   D. convenience.

9. In determining the total number of elevators required in a new building, an architect should consider the

   I.   total number of floors.
   II. estimated volume of traffic.
   III. number of hours of operation.
   IV. elevator roping system.
   V.   elevator car size.

   A. II and V
   B. I, III, and IV
   C. I, II, and V
   D. II, III, IV, and V

10. Escalators are most effectively used in situations that require

    A. low cost.
    B. high speed.
    C. considerable life safety.
    D. considerable user capacity.

FIRE SAFETY

INTRODUCTION

Fire protection in buildings is the subject of this lesson, and two different codes are the source of most of the information. The Uniform Building Code (UBC) is a model building code that specifies the fire protection that must be incorporated into any building. The features that are covered include fire ratings of walls, floors, and roofs; and the requirements for sprinklers, exits, standpipes, etc.

The 16 volumes of the National Fire Code specify details of fire protection systems, particularly Volumes 10 through 13. In fact, they go beyond the purview of the architect in detailing how a building is to be used and how equipment is to be maintained.

While there are other fire protection codes, this lesson is based on these codes, since they are referred to by the NCARB. The candidate should have access to a copy of the current UBC as he or she studies this lesson.

FIRE SAFETY PRIORITIES

Traditionally, fire protection codes have had three goals. The first is to afford protection or escape for the occupants of a building. This may be accomplished by either of two means: evacuation from the building (egress), or moving the occupants from an endangered area to a protected area elsewhere in the building, a *place of refuge*. In small structures, evacuation is the usual approach, but in high-rise structures a place of refuge is the only practical method.

The second aim of fire protection is to insure sufficient structural integrity in the building so that fire fighters may enter and fight the fire without excessive risk of being trapped or injured by collapsing portions of the building.

The third and least significant fire protection goal is to allow the building to survive a fire so that it may be economically restored afterwards.

A fourth goal that is becoming a major thrust of building codes is to prevent fires from starting, or once they have started, to extinguish them immediately and automatically. In the MGM Grand Hotel fire in Las Vegas, parts of the casino and hotel were completely destroyed, and there was significant loss of life. Yet immediately adjacent to the point of origin of the fire, a restaurant that was sprinklered survived the fire with tablecloths intact.

BUILDING CODE

In order to determine the UBC requirements for a specific structure, the architect must classify the building by its function, its construction type, and its location.

Occupancy

To determine the degree of fire resistance required in a given structure, the UBC has established a number of specific occupancy classifications. These are listed in Table 3-A of the UBC, which the candidate should review. Generally, however, these occupancies may be categorized as follows:

For basic classification, all buildings are assigned a *group* letter, and then for more specific definition, a number that identifies the subcategory or *division* within the basic

TABLE 10.1 – OCCUPANCIES	
Group	**Description of Occupancy**
A	Assembly
B	Business
E	Educational
F	Factory and Industrial
H	Hazardous
I	Institutional
M	Mercantile
R	Residential
S	Storage
U	Utility

classification. Open parking garages, for example, are Group S Division 4, or S-4 occupancy.

Construction Type

Buildings are also categorized by their type of construction that, in turn, determines their degree of fire resistance. The candidate should be aware of the requirements for the major building elements for each type of construction, which are specified in Table 6-A of the UBC.

The basic method of assigning fire resistance ratings is to test each construction material or assembly under standardized conditions for a period of time, typically one to four hours. The detailed requirements for each type of construction are specified in Chapter 6 of the UBC, but the candidate need not study this chapter in depth, skimming should be sufficient. The basic idea is that there are five construction types, ranging from the most fire-resistive *(Type I)*, to conventional wood stud construction *(Type V)*.

The location of a building on its property with regard to building setbacks, alleys, public streets, and property lines also affects the required fire resistance ratings of exterior walls. In this regard, the candidate should review UBC

Table 5-A. The basic intent of the requirements is that fire should not be allowed to spread from one building to the next, so increased fire resistance is required at property lines, etc. This gives the local fire department time to put out the fire in a single building before it consumes an entire block of buildings. For the same reason, openings in exterior walls are limited in size, so that fire cannot easily pass through the rated walls.

The maximum floor area of a building is limited in accordance with Table 5-B of the UBC. Buildings of more fire-resistive construction are allowed greater floor area. In fact, buildings of Type I construction are permitted unlimited floor area for most occupancy groups. It is also important to note that the installation of an automatic sprinkler system in an occupancy which does not otherwise require it permits a very significant increase in allowable floor area.

Similarly, the height and number of stories in a building are limited for various types of construction, depending on whether or not the building is sprinklered.

The number of occupants assumed to be in a building is based upon the building's occupancy and floor area.

To summarize, a building must comply with the detailed requirements of the code that specify fire-resistive ratings for walls and other elements, height, and exits for the different occupancies and types of construction.

COMPARTMENTATION

Frequently, more than one occupancy group is located within a given structure. In order to permit such mixed occupancies, the construction of the walls separating them is also specified. For example, if a retail store (M occupancy) is in the same building as a repair garage (S-3 or H-4 occupancy), a one-hour separation is required between the two, in accordance with Table 3-B of the UBC.

Similarly, when a design requires more floor area than is permitted for its occupancy, the space may be separated into two or more portions, each of which must comply with exiting and other requirements as if it were a separate building. The walls and floors separating the compartments must then have a fire rating of four hours or two hours, depending on the building type.

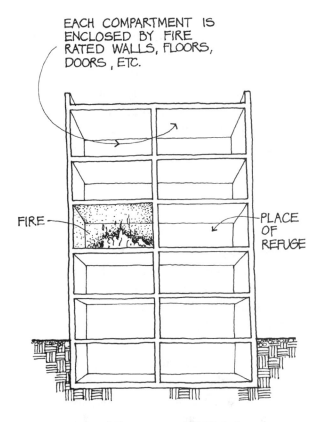

EACH COMPARTMENT IS ENCLOSED BY FIRE RATED WALLS, FLOORS, DOORS, ETC.

FIRE

PLACE OF REFUGE

COMPARTMENTATION

In order for the compartmentation to be effective for fire safety, all openings through the walls must be closed with fire rated devices. This includes not only doorways and windows, but also air ducts that pass through the fire separations. In order to have a fire rating, a prototype of the entire fire assembly must have been previously tested and approved by Underwriters' Laboratories (UL) or a similar, recognized testing agency. This test is done by burning a specified fire on one side of the assembly for the specified time and then testing its function. With doors, for example, this means shooting a stream of water from a fire hose at it to see if it pops out of the frame. Note that not just the damper or the door must be fire rated, but the entire assembly—including the frame, the mounting hardware, the latches, the hinges, the sill, and the door itself. In addition, fire assemblies must be either self-closing or automatic if subject to an increase in temperature or to products of combustion. Self-closing assemblies are typically held open by means of fusible links which melt when the ambient air temperature exceeds a certain value, often 165° F. The assembly will then close and latch by means of a spring or gravity device.

EXITS

The code requires that exit passageways be provided in every building from every part of every floor to a public street or alley. This means having at least one exit from every building or compartment of a building, and when the rated number of occupants exceeds a certain number, two or more exits are required. For example, assume that a dance studio occupies a 40 × 50 foot space on the second floor of a store building that has separated stairways to the street. Its occupancy is A-3, and it is assigned a maximum capacity of (40 × 50)/7 = 286 persons (see Table 10-A). Since any dance floor with more than 50 persons requires two exits, this space requires two legal exits. Most buildings require *two or more exits*. The reasoning is that if there were

only one exit, it could be blocked by the fire and the occupants would be trapped. If there are two means of egress, it is unlikely that both would blocked by fire.

Each exit stairway must be within 150 feet of any point (200 feet in a sprinklered building). These distances are increased to 200 feet and 250 feet respectively in the 1997 UBC. Thus, a 300-foot-long building requires at least three exit stairways (one in the middle and one at each end).

Since an exit passage may be the escape route for several spaces, the total flow in the passage must be considered. This is done by determining the total occupant load to be served by the passage and multiplying by 0.2. The result is the minimum exit width in inches. Under no circumstances should any part of an exit passage be less than *44 inches wide.*

All of the doors in the exit passage must swing in the direction of travel, and must be unlatched or activated by *panic hardware.* Panic hardware is a door-latching assembly that will release the latch if a force not exceeding 15 pounds is applied to it, such as would occur if someone stumbles blindly into it in the direction of travel. The assumption is that the hallway may be filled with smoke, and the occupants may not be able to see (or think) properly.

Exits from all public spaces must be provided for handicapped persons. This may mean ramps or paths to safe compartments. Elevators may only be used if they are not connected to heat responsive call buttons, for obvious reasons. In general, elevator shafts collect smoke, somewhat like a chimney, from the floor that is on fire, and their use should be avoided if at all possible.

All exit stairways must be of fire-resistive construction. Stair enclosure walls in buildings which are four stories and greater in height, or of Types I and II-F.R. construction, require a minimum of two-hour construction. One-hour rated walls are acceptable for all other buildings. In buildings greater than 75 feet in height, the entire required exit enclosure must be pressurized, with pressurization occurring automatically upon activation of the fire alarm system.

Large floors should be subdivided so that the first means of escape is simply to get across the division to the other section of the same floor. This limits the spread of fire, so that fire fighting personnel and equipment can more easily control the fire, and is a more satisfactory means of egress for handicapped individuals than moving to another floor. Again, all doors and dampers in the division boundary wall must be fire-rated.

CLASSES OF FIRES

Fires have been divided into four classes: A, B, C, and D. Class A fires are those involving ordinary materials, including wood, paper, cloth, and rubber. These can generally be extinguished with water, or with an extinguisher that sprays water. Class B fires involve flammable gases and liquids, such as natural gas, gasoline, oil, etc. These tend to float on top of water, making water ineffective in extinguishing them. Class C fires involve electrical equipment, and the extinguishing medium must be electrically non-conductive, and therefore water is not acceptable. After the source of electricity is disconnected, Class A or B extinguishers may be used. Class D fires involve combustible metals such as sodium, potassium, magnesium, etc., and require special extinguishers. In fact, sodium at room temperature may actually burst into flames on contact with water.

Special Extinguishing Media

There are several special media used in hand-held and automatic fire extinguishing systems. One such medium is Halon (Halon 1301 or Halon 1211), which is not toxic for brief exposures, and which may be used safely on Class B and C fires. However, prolonged exposures should be avoided. It displaces oxygen, which is useful in fighting fires, but it can eventually result in asphyxiation. Carbon dioxide (CO_2) also displaces oxygen. When either of these is used in an automatic system, locally audible and visible alarms must be provided that warn personnel to leave the vicinity. Both of these extinguishing media are preferred in areas where there are valuable documents or artwork. Halon is particularly common in computer installations. Neither equipment nor records are damaged, but the fire is smothered. The cause of the fire should be found immediately, however, as it is likely to restart when oxygen returns.

FIRE DETECTION

There are three forms of fire detection available. They are based on ionization, photoelectric detection, or temperature sensing.

Ion Detectors

The *ionization detector* responds to the chemical *products of combustion (POC)* present in the air during a fire, even in the earliest stages. These may be visible or invisible; the ionization detector is sensitive to both. Such detectors are now inexpensive and available even for home use. Batteries should be checked periodically. The one problem with ion detectors is that they also detect smoke from the kitchen or cigarette smoker.

Photoelectric Sensors

The *photoelectric detector* reacts to visible smoke in the air that blocks a beam of light. This may measure across a larger volume of air, but may also miss some of the early signs that the ion detectors will pick up. Given the low cost of ion detectors, they have surpassed the photoelectric detectors in popularity.

Both systems are actually preferable to waiting for human detection of fire. They are capable of sensing smoldering fires long before the fire is visible to the naked eye, which tends to be too late with many kinds of fire.

Small or smoldering fires are particularly dangerous because of a phenomenon known as *flashover*. Smoldering fires release gases that are at fairly high temperatures, and that collect near the ceiling. The ceiling materials become extremely hot over a rather broad area. When they finally reach combustion temperatures, they tend to do so all at once, and a very small fire can become a huge one in moments. In extreme cases, the gases then superheat, almost exploding out of the area where the fire started.

All detectors should be placed close to or on the ceiling. Vertical circulation spaces in the building often provide good locations. For example, smoke created in one portion of a home will often rise up a stairway.

Heat Actuated Sensors

A less sensitive detector is the heat actuated type. There are several ways this is done, from the original simple fusible link to more sophisticated electronic devices. At the simplest, a piece of wax or paraffin separates the contacts of an alarm circuit, and when it becomes warm and melts, the alarm goes off. Similarly, fire separation doors in older buildings were spring loaded to shut, or to drop, and the only thing holding them open was a piece of wax. When the wax melted, the doors shut. Although primitive, this is still surprisingly effective, and has saved numerous buildings, especially some of our historical landmarks.

Such devices rarely cause false alarms, but often are actuated too late to save the room in which the fire began.

Several other functions may be set in motion by the fire alarm. These may include:

1. Actuating remote alarms, in the building or even at the local fire station.
2. Actuating other extinguishing systems.
3. Overriding elevator controls.
4. Closing fire doors, fire dampers, and otherwise controlling smoke and fire migration.
5. Varying fan speeds throughout the mechanical equipment system.

Modern buildings are often equipped with fire fighting stations, which become active in case of a major fire. Elevators and intercom systems may be controlled from the station. Some are provided with information from the fire detection devices as well, so that the fire fighters may know the progress of the fire throughout the building at a glance.

STANDPIPES

The normal water distribution system in a building is rarely adequate to fight a fire within the building. For this reason, *standpipes* are used. These are not intended to go to each space, but rather to distribute large volumes of water to each floor, from which the fire fighting hoses and equipment can distribute the water to the spaces where it is needed. There are two types of standpipes, dry and wet.

Dry Standpipes

Dry standpipes are large diameter water risers that are normally empty and not connected to a water supply. The lower end terminates at street level where the fire department can conveniently connect it to a fireplug via a

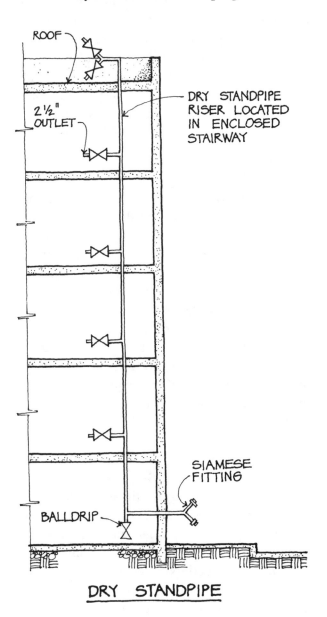

DRY STANDPIPE

pumper truck, which is capable of pumping the water up through the pipe.

The fitting at the lower end of the standpipe is referred to as a *siamese fitting* and is arranged to accept either two- or four-hose connections from fire department pumpers, depending on whether the dry standpipe is four or six inches in diameter. A 2 1/2 inch outlet connection must be provided at every floor level higher than the first floor and at the roof. If the dry standpipe extends more than 75 feet above grade, the pipe connections must be provided in every required stairway as well.

The dry standpipe is, in effect, a portion of the fire department's equipment, which has been permanently installed in the building. To use it, the fire fighters connect a pumper to the siamese fitting at the lower end, and then carry their own rolls of hose up inside the building, connecting them to the 2 1/2 inch outlets wherever necessary.

One of the benefits of the dry standpipe is that there is no need to worry about rusting or freezing. To insure that it remains dry, an automatic drain valve (called a *ball drip*) is located at the lowest point of the system.

Wet Standpipes

Wet standpipes are required in buildings of four or more stories in height, throughout most theaters and other places of assembly, throughout all hazardous occupancies, and throughout all Group I, B, S, and M occupancies. Wet standpipes are provided primarily for the use of the occupants of the building, but they must also be equipped with siamese fittings so that the fire department may supply additional pressure and flow. Wet standpipes must be located in a building so that every point of every floor is within 30 feet of the end of a 100-foot hose attached to an outlet. These hoses are usually pre-attached and stored folded in a wall case with a glass panel making them clearly visible.

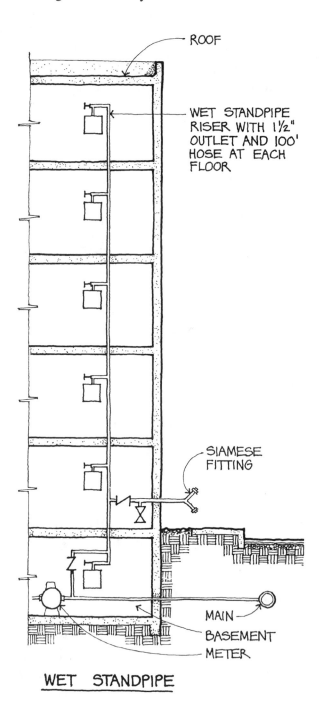

ROOF

WET STANDPIPE RISER WITH 1½" OUTLET AND 100' HOSE AT EACH FLOOR

SIAMESE FITTING

MAIN
BASEMENT
METER

WET STANDPIPE

The wet standpipe system must be designed to supply at least 35 gpm at 25 psi minimum for at least 30 minutes. The water supply system itself must be adequate to provide 70 gpm for 30 minutes at 25 psi minimum. The water supply may be a pressure tank, a gravity tank, or an automatic pump as long as the power source for the pump is safe.

Buildings that exceed 150 feet in height require a *combination standpipe* for every stairway that extends from ground to roof, which replaces the pipe otherwise required. The combination standpipe must be equipped with the 2½ inch outlets for the fire department and also with 1½ inch hose racks as in a wet standpipe system.

SPRINKLER SYSTEMS

Automatic sprinklers are a widely-used and very effective means of extinguishing, or at least controlling, fires in their early stage. Being automatic, they do not require the attendance of a person for operation, thus providing protection even when a structure is not occupied. The actual flow of water from a sprinkler head is comparable to a very heavy rain storm, and therefore is not sufficient to cause injury to human occupants. However, sprinkler systems can damage a building's contents, such as paper, records, electronic equipment, artwork, or merchandise. Flow rates range from 5 to 20 inches of water per hour, depending upon code requirements.

The UBC requires automatic sprinklers in many occupancies and encourages them in others. With some exceptions, sprinklers are required in all of the following occupancies: basements and cellars of all buildings, except private houses and garages; the backstage area, dressing rooms, workshops, and storage areas of theaters; and any concealed spaces above stairways in schools, hospitals, institutions such as prisons, and places of assembly such as theaters and arenas. Automatic sprinklers are also required over all rubbish and linen chutes, except in private dwellings; in retail sales areas over 12,000 square feet per floor or over 24,000 square feet gross; and in all places of assembly over 12,000 square feet in area. The code permits an increase in allowable floor area and height, and allows wider spacing of exit stairs for sprinklered buildings. In many jurisdictions, the requirements for automatic sprinklers are more stringent than those in the UBC.

Wet and Dry Systems

The simplest type of sprinkler system consists of a pattern of sprinkler heads, each of which is equipped with a fusible plug or fusible link. In the event of a fire or very high ambient temperature, the fusible plug will melt and the water pressure in the pipe will cause a spray of water through the sprinkler head. Some heads are inset into the ceiling to be less conspicuous. Such heads often also pop out when activated to give better distribution. As with standpipes, there are both wet and dry systems.

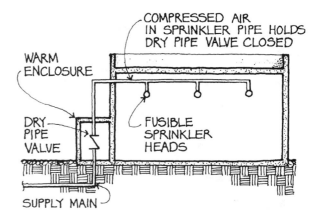

DRY PIPE SPRINKLER SYSTEM

The wet pipe system has the advantages of quick response and low initial cost. However, its disadvantages include the possibility of freezing and unnecessary wetting of building contents. To overcome the freezing problem, the dry pipe system was developed. In this system, the sprinkler piping between the dry pipe valve and all of the sprinkler heads is empty of water and is filled with compressed air. The pipe valve may be located in a warm enclosure. The disadvantage is that when the sprinkler is first activated, nothing but compressed air comes through it until the system between the valve and the sprinkler has been flushed of air. This may cause a dangerous time delay in long pipe runs.

Preaction System

The preaction system is a variation of the dry system that requires that both the sprinkler head be activated and an independent fire sensing device triggered. This avoids accidental discharge of the sprinklers and resultant water damage. It is not as fail-safe as either the wet or dry system, however. All of the above systems cause only the activated sprinkler heads to release water.

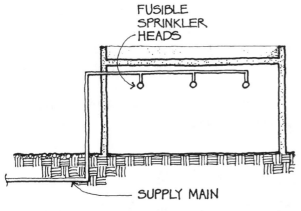

WET PIPE SPRINKLER SYSTEM

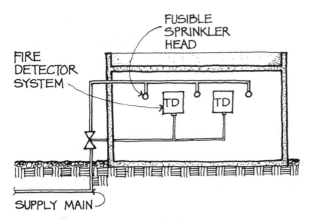

PREACTION SPRINKLER SYSTEM

Deluge System

The *deluge system* is based on the idea that if there is a fire somewhere within the space, wetting the entire space is the safest course of action. This is generally reserved for areas of high fire hazard. In the deluge system all of the sprinkler heads are wide open at all times, but the pipe system is empty of water. Release of the water is actuated by a heat or fire detection system installed in the area to be protected, which activates a valve, flooding the system with water.

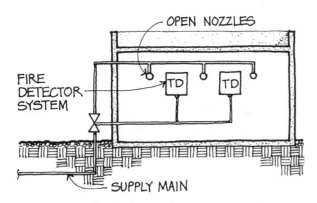

DELUGE SPRINKLER SYSTEM

All sprinkler systems must also have a *siamese connection* outside the building to permit the fire department to augment the overall flow.

Hazard Levels

In order to determine how great an area may be served by one sprinkler head, three major hazard levels have been established. *Light hazard* identifies areas where the quantity of combustible materials is relatively low. Typical light hazard occupancies include churches, hospitals, museums, offices, and residential occupancies. A greater degree of hazard is identified as *ordinary hazard*, which is subdivided into groups 1, 2, and 3, with progressively greater hazard as the group number increases. Group 1, for example, includes automobile garages and laundries. Group 2 includes large stack room areas of libraries and printing and publishing plants. Group 3 includes paper processing plants and tire manufacturing plants. *Extra hazard* is the greatest degree of hazard and applies to areas such as aircraft hangars and explosives handling areas.

Sprinkler heads *must never be repainted.* This ruins the temperature sensing capability of the fusible link, and may cause the mechanism to be jammed by the paint.

Insurance Company Requirements

Because insurance companies base their rates, in part, on the adequacy of the sprinkler system in an insured structure, they always require, as a condition of the insurance contract, that the owner immediately notify them of any proposed or actual changes in the sprinkler system protecting the structure. Failure to do so may result in loss of coverage.

SUMMARY

We have discussed the basic principles of fire safety in buildings. There are several strategies, which fall into two groups: (1) fire-resistive construction and safe egress, and (2) preplaced firefighting equipment. Fire-resistive construction involves occupancy groups, construction types, fire-resistive ratings of floors and walls, exiting requirements, and maximum allowable floor areas and heights. Preplaced fire-fighting equip- ment includes fire detection equipment, extin- guishers, standpipes, and sprinkler systems. As always, understanding the principles is better than memorizing tables.

LESSON 10 QUIZ

1. Compartmentation refers to
   A. sealing special chemicals in compart-ments, which are released when fire is detected.
   B. containing the fire in the area in which it began.
   C. dividing a sprinkler system into com-partments based on required flow rate.
   D. dividing the city into Fire Zones 1, 2, and 3.

2. The minimum exit corridor width is
   A. 36 inches.   C. 44 inches.
   B. 39 inches.   D. 48 inches.

3. The maximum distance from a room to a fire stair exit is
   A. 75 feet.   C. 150 feet.
   B. 130 feet.   D. 200 feet.

4. If a wet standpipe system is installed, every point on every floor must be within how many feet of a connection?
   A. 75 feet   C. 150 feet
   B. 130 feet   D. 200 feet

5. Flashover occurs when
   A. all the materials reach combustion tem-perature at once.
   B. one building's roof ignites the next building's roof.
   C. a fire thought to be extinguished restarts itself from the embers.
   D. high temperatures trigger a sprinkler head.

6. Building occupancy groups are determined by
   A. the construction materials used in the structure.
   B. the number of people expected to use the building.
   C. the function for which the building is designed.
   D. the size of the building.

7. Dry standpipes are
   A. used by building occupants to put out a fire before it gets out of control.
   B. filled with Halon gas or carbon dioxide, and triggered automatically.
   C. used by fire fighters, who have to pump water into them.
   D. used to drain floors of excess water dur-ing fire fighting.

8. The major benefit of using sprinkler systems is that
   A. they do not require that someone be in attendance to be effective.
   B. they are considerably less expensive than standpipe systems.
   C. they do not damage valuable artwork, books or records.
   D. they may also be used as domestic plumbing systems.

9. Which of the following fire detection
   devices will sense the products of
   combustion, whether or not the fire is very
   large?

   A. Ionization detector

   B. Smoke detector

   C. Sprinkler head with fusible link

   D. Heat detector

10. Which of the following are required to have
    a siamese connection at the base?

    I.   Wet standpipes

    II. Dry standpipes

    III. Wet sprinkler systems

    A. I and II   **C.** I and III

    B. II and III   **D.** I, II, and III

The following glossary defines a number of mechanical and electrical terms, many of which have appeared on past exams. While this list is by no means complete, it comprises much of the terminology with which candidates should be familiar. You are therefore encouraged to review these definitions as part of your preparation for the exam.

Absorption Coefficient The percentage of incoming energy that is absorbed. In measuring radiant energy (light or heat) it is a unitless ratio that may vary depending on wavelength. In acoustics, it is called a sabin and may vary depending on frequency.

Absorption Refrigeration A cooling process resulting from the absorption of vapor by a brine solution that is then heated to remove the moisture. The heat may be supplied by solar or other heating sources.

AC The abbreviation for either air conditioning or alternating current, depending on the context.

Acoustic Power Level See Power Level.

Air Gap An unobstructed vertical path, open to the atmosphere, separating the outlet of a faucet from the overflow rim of the fixture it serves. The purpose is to prevent a momentary vacuum in the supply pipe from siphoning water from the fixture back into the supply pipe.

Alternating Current An electric current that reverses its direction at regular intervals, generally 60 times per second in the United States. A plot of the voltage over time is a sine wave.

Ambient Relating to a general or all surrounding condition. In thermal processes, it refers to the air temperature, as distinct from that of surfaces or objects. In lighting, it refers to the background light level, and in acoustics, it refers to the background noise level.

ASHRAE The abbreviation for the American Society of Heating, Refrigerating, and Air Conditioning Engineers, the source of most of the standardized information on the subject.

Aspect Ratio The ratio of the longer to shorter dimension of an air conditioning duct, that affects duct friction, or of a room, that relates to light reflection.

Ball Drip The automatic drain valve at the base of a dry standpipe.

Blow Down The drain that removes dirt that builds up in the floor pool of an evaporative cooling tower.

BTU The abbreviation for British Thermal Unit, a unit of heat energy, which is defined as the amount of heat required to raise the temperature of one pound of water by one degree Fahrenheit.

BTUH The abbreviation for Btu's per hour, an energy flow rate.

Building Automation The control by automatic equipment of many functions in a large building, usually including the HVAC system, the fire detection and alarm system, and building security.

CFM The abbreviation for cubic feet per minute, which is the flow rate of air in a mechanical system or duct.

Chiller A piece of equipment that cools water for use in an air conditioning system.

Chill Factor A fictitious temperature assigned to a combination of actual temperature and wind velocity that has the same physiological effect as still air at the chill factor temperature.

Chlorination The addition of small amounts of chlorine to a water source to kill bacteria.

Circuit Vent In sanitary drainage, any vent that serves two or more traps.

Clerestory A window above eye level that admits daylight, such as the high windows in a cathedral.

Code An organized body of rules and regulations adopted and enforced by a governmental unit, such as a building code.

Coefficient of Performance The ratio of the amount of heat energy delivered by a heat pump to the amount of energy supplied, or the ratio of the amount of heat energy removed by a refrigeration machine to the amount of energy expended in its removal. It is similar to efficiency, but often exceeds 100 percent, and is therefore expressed as a number, i.e., 2.7, which is typical for a heat pump.

Coefficient of Utilization The ratio of useful light arriving at the work plane to the amount of light emitted by the source. The CU depends on the reflectivity of different surfaces and the aspect ratios of the ceiling, wall, and floor cavities.

Comfort Zone The combination of thermal and environmental conditions within which a human is comfortable, often shown on a psychrometric chart.

Conductance (C) The rate at which a specific thickness of a given material conducts heat.

Convection The heat transfer process that occurs when a warm fluid rises, displacing cold fluid which then falls.

Cycle One complete set of repeating events, typically used with alternating current or sound.

Cycles per Second (CPS) A measure of frequency in electric current or acoustics, i.e., the number of cycles per second of a wave or oscillation. In acoustics, the term has been largely replaced by Hertz, where 1 cps = 1 Hz.

Daylighting The use of natural light from outside to replace or augment electrical light outdoors, which produces energy savings.

dBA A decibel measured in the A scale, that is weighted to account for the special sensitivities of the human ear.

Decibel (dB) A logarithmic measure of sound intensity level.

Degree Day (DD) The amount by which the average outdoor temperature at a particular location is below 65°F for one day. Degree days may also be summed and stated for a month or a year.

Delta Connection A method of connecting windings on a three phase transformer, end to end, that results in a triangular shape.

Dew Point Temperature The temperature of air at that the water contained in the air begins to condense and form dew. The dew point for a given air sample is always lower than or equal to its current temperature.

Diffuser A device through which the air from a duct enters a room, or a device through which the light from a fixture enters a room.

Dry Bulb (DB) The temperature of air as read on an ordinary glass thermometer.

Dry Pipe Sprinkler A sprinkler system whose pipes are normally pressurized with only air, thus being invulnerable to freezing temperatures. Upon actuation, the air is vented and supply pressure forces water through the system.

Dry Standpipe See Standpipe.

Economizer Cycle An energy-saving strategy in which a part of the HVAC system is shut off while the rest is used, such as shutting off the refrigeration when the outside air temperature is low, while the fan continues to operate.

Effective Temperature A fictitious temperature having the same physiological effect as air of a standardized temperature, humidity, and velocity.

Efficacy The ratio of the lumens emitted by a lamp to the electrical power consumed by the lamp.

Emissivity A factor that represents the rate at which a given surface material gives off or emits radiant energy. The emissivity varies from 0 to 1.0, where 1.0 is the theoretical emissivity of a perfect black box at the same temperature.

Enthalpy The total of sensible plus latent heat stored in the air. It is also known as Total Heat.

Exit A continuous and unobstructed means of egress to a public way. Its minimum width is generally 44 inches.

Fire Assembly A complete fire-resistive assembly consisting of a fire door, fire damper, or fire window and its mounting frame and hardware. The entire assembly, not just its components, must be approved and labeled by a testing agency that inspects the materials and workmanship during fabrication at the factory. Available ratings are 3/4, 1, 1-1/2 and 3 hours.

Fire Door See Fire Assembly.

Fixture Unit A unit of liquid flow used in sizing both supply and drainage pipes.

Flame Spread Rating A numerical classification indicating the rate at which flame will spread in or on a given material, in which higher numbers flame up more rapidly.

Flushometer Valve A valve that releases a definite amount of water into a plumbing fixture each time it is actuated.

Footcandle The basic unit of illumination arriving at a work plane. One footcandle is equal to one lumen per square foot.

Forced Air System A heating or cooling system that uses a fan to circulate heated or cooled air through ducts to the occupied spaces.

Forced Convection The movement of a fluid by a fan or a pump, in order to force heat exchange.

Four Pipe System A hot and chilled water system having separate return lines for each supply line, and with no mixing of the two streams.

Frequency The number of cycles that occur per second, either in alternating current or acoustics. In acoustics, the frequency determines the pitch.

Frost Line In a given location, the maximum depth in soil that is expected to freeze in cold weather. Water piping must ordinarily be buried below the frost line to protect against freezing.

Fusible Link A piece of wax or paraffin that melts at a predetermined temperature, setting off a sprinkler head, an alarm system, or otherwise actuating a fire protection device.

Globe Thermometer A thermometer that measures Mean Radiant Temperature (MRT).

Ground An electrical conductor connected to the earth or to a pipe extending into the ground, used to dissipate hazardous currents into the earth.

Halon A gaseous fire extinguishing medium that smothers fires, often used in automatic systems in computer rooms.

Heat Pump A refrigeration loop used to bring heat into a space instead of removing heat from it. The term is also used for an entire system of such units attached to a recirculating heat sink.

HEPA Filter A high efficiency particulate air filter, which removes dust and other tiny particles from a moving air stream.

Hertz (HZ) The frequency of a sound, or of an alternating current, equal to the number of cycles per second.

High Intensity Discharge (HID) A family of lamps consisting of a quartz envelope inside a glass envelope. The inner quartz tube can stand higher temperatures, and allows for the current to arc between the two electrodes exciting a plasma of mercury, metal halide, or high pressure sodium.

HP or BHP Horsepower or brake horsepower, a unit of power, equal to roughly 746 watts.

HVAC The abbreviation for heating, ventilation, and air conditioning.

Illumination The intensity of light falling on a surface, usually expressed in footcandles.

Impact Isolation Class (IIC) A rating of the degree of isolation of a floor against the transmission of impact noises.

Infiltration The leakage of air through cracks around windows and other building elements.

Intensity Level The intensity of sound at a given location, measured in watts per square meter, or more commonly in dB where the reference level is 10^{-12} watts/meter^2 or 10^{-16} watts/cm^2.

Inverse Square Law A physical principle that states that the intensity of a phenomenon is inversely proportional to the square of the distance from the source to the measuring device. It holds true for point sources of light, and for sound in an open field.

Invert The lowest point of the inside of a drain, pipe, channel, or other liquid-carrying conduit.

Ion Exchange A process of water softening in which calcium and magnesium ions are replaced by sodium ions. This process is also known as the zeolite process.

Ionization Detector   A fire detector that detects the products of combustion (POC) even before they are visible to the naked eye.

K Factor   The thermal conductivity of one square foot of a material per inch of thickness, with a surface temperature difference of one degree F.

Kilowatt (KW)   A unit of electric power, equal to 1,000 watts.

KVA   A rating for transformers equal to the product of volts and amperes divided by 1,000. The product of the KVA and the power factor gives the power in kilowatts.

Latent Heat   The heat added to or removed from a substance when it changes its state. See Sensible Heat.

Light Shelf   An overhang, either outside or inside or both, that is used with a clerestory to reflect light up onto the ceiling, and reduce direct light adjacent to the window below.

Lumen   A unit of light, defined as the amount of light passing through one square foot at a distance of one foot from a one candlepower source.

Luminaire   A complete light fixture, including lamps.

Mass Law   The theoretical law that states that for each doubling of mass in a wall, there is a 6 dB drop in the amount of sound transmitted.In actual practice, it is usually closer to 5 dB.

Mean Radiant Temperature (MRT)   The weighted average of all of the temperatures of all of the surfaces visible from a given position.

NC Curves   A single-number system for specifying a maximum SPL level in a given location, using standardized reference contours. The curves weight the frequencies to which the human ear is sensitive.

NEC   The abbreviation for National Electrical Code.

Neutral   The wire or conductor in an electrical system that is equidistant in voltage from the phase conductors of the system. It is not the same as ground.

Noise Reduction Coefficient (NRC)   A one-number rating system giving the average sound absorption coefficient of a material at frequencies of 250, 500, 1,000 and 2,000 Hz.

Occupancy   The purpose for which a building is intended to be used.

Occupancy Group   A designation for a group of several occupancies that have comparable fire safety considerations, and which are therefore grouped together by the code.

OSHA   The abbreviation for the Occupational Safety and Health Act, which regulates working conditions.

Passive Solar Design   The practice of orienting and sizing a building, its windows, and its internal masses in such a way that it responds to the sun and to the climate, without the use of mechanical equipment.

Perm   The unit of permeability for a given material, expressing the resistance of the material to the penetration of water or water vapor through it. One perm is equal to the flow of one grain of water vapor through one square foot of surface area per hour with a pressure difference of one inch of mercury.

Permeability   The property of permitting passage of water or water vapor through a material without causing rupture or displacement.

Potable Water   Water that is suitable for drinking.

Power Factor (PF)   In an electrical circuit, the ratio of real power in watts to the product of voltage and current.

Power Level (PWL)   The logarithmic expression for the acoustical power at the source of a sound. It is also known as Acoustic Power Level.

PPM   The abbreviation for parts per million.

Psychrometric Chart   A graph showing the relationships between temperature, humidity, relative humidity, and enthalpy.

Reheat   The adding of sensible heat to a supply air stream that has been previously cooled.

Relative Humidity (RH)   The ratio of the moisture content of the air to the maximum possible content at the same temperature.

Reverberation   The persistence of sound in an enclosed space after the source has stopped.

Reverberation Time   The time it takes a 60 dB sound to completely die away in a closed room after the source has stopped.

Sabin   The unit of sound absorption equivalent to the absorption of one square foot of open window.

Sensible Heat   Heat that changes the temperature of a substance, and does not represent the addition of any moisture to the substance, or any change of state. See Latent Heat.

Shading Coefficient (SC)   The ratio of the solar heat gained through a window with shading devices to the solar heat gained by a single pane double strength clear glass window. Shading devices, such as Venetian blinds, lower the SC.

Siamese Fitting   A Y-shaped hose attachment at the base of a building, which allows the fire department to connect a fire hydrant through a pumper truck to provide or augment water flow to a standpipe.

Smoke Developed Rating   A numerical rating derived from a standardized fire test procedure. Larger numbers indicate a greater density of smoke.

Smokeproof Enclosure   A continuous enclosed stairway separated from the building at each floor by an open vestibule that allows smoke to vent away without entering the stair.

Soil   A sanitary drainage term referring to the waste from urinals, water closets, and fixtures of similar function.

Sone   A subjective system of measuring loudness, based on the reference point of one sone equal to a sound pressure level of 40 dB.

Sound Level Meter   A meter that measures the sound pressure level and gives a reading in dB.

Sound Pressure Level (SPL)   The logarithmic expression of the pressure exerted by sound waves on the receiver. The reference pressure is 2×10^{-5} newtons per square meter.

Sound Transmission Class (STC)   A single-number rating for the evaluation of a particular construction cross-section in terms of its transmission of airborne sound. The higher the STC rating, the more effective the construction is at stopping airborne sound.

Sprinkler System   A system used to extinguish fires automatically by releasing water or other substances. See Dry Pipe Sprinkler and Wet Pipe Sprinkler.

Stack Vent   The portion of a soil or waste stack that is above the highest branch drain connected to the stack. Its sole function is to vent to the outside air.

Standpipe   A vertical supply pipe for firefighting. Dry standpipes are empty, and must be connected to a fire hydrant via a siamese connection and a pumper truck. Wet standpipes are pressurized and filled with water, to serve attached hoses within the building, on each floor. Wet standpipes also have siamese connections to allow the water flow to be augmented from fire hydrants.

Star Connection   A method of connecting the windings on a three-phase transformer in which one end of all three windings is connected to a common neutral center point, forming a Y shape. Same as Wye Connection.

Starter   A device that starts the arc in a neon or fluorescent lamp, or a contactor and overload relay used in starting some electric motors.

Steam Trap   A valve that permits passage of air or water, but not steam, often used with steam radiators.

Sweating   The method of soldering copper plumbing, or the condensation of water on cold pipes or building materials.

Thermosiphon   The method of using a heated surface and the resulting convection to move a fluid out of a space. In solar hot water heaters, the collector is below the storage tank, and the water is circulated automatically by convection when it is heated. Thermosiphoning may also be used to ventilate a building by sending the warmed air out the top, and siphoning in cooler air at the bottom.

Three Pipe System   A hot and chilled water system having a common return pipe for both supply lines.

Ton   The amount of cooling required to create a ton of ice in a 24-hour period, equal to a steady rate of 12,000 Btuh.

Total Heat   See Enthalpy.

Transmission Loss (TL)   The reduction of sound that occurs when a given wall transmits sound from one room to an adjacent room, expressed in decibels.

Trombe Wall   A form of mass wall that transfers heat by causing convection into the room behind it, as well as conduction.

Two Pipe System   A hot or chilled water system having only a supply and return line. It can supply only heated or chilled water to a zone, but not both simultaneously.

UBC   The abbreviation for Uniform Building Code, one of the most widely-adopted model building codes in the United States.

U Factor or U Value   The thermal conductivity of a particular wall section, expressed in Btu's per hour per degree Fahrenheit per square foot.

Vacuum Breaker   An automatic valve that admits air into a supply pipe rather than allowing the pipe to suction or siphon polluted water back into the supply.

Variable Air Volume (VAV)   An air conditioning system that accommodates thermal load changes by varying the flow of supply air into a conditioned space instead of varying the temperature of the air.

Vent Stack   A vertical pipe that vents several sanitary drainage lines, in order to break the siphoning suction that would occur when water drops through the system.

Watt   The basic unit of electrical power, equal to the product of volts and amperes in direct current systems, equal to 3.41 Btuh.

Wavelength   The length of one complete cycle or waveform, for light or sound waves. In light, the dominant wavelength determines the perceived color.

Wet Bulb Temperature (WB)   The temperature attained by a glass thermometer whose bulb is covered with a wet sock and placed in an air stream moving at 1,000 feet per minute.

Wet Pipe Sprinkler   A sprinkler system that is continually pressurized with water. If a fusible sprinkler opens, water is immediately forced through the sprinkler head.

Wet Standpipe   See Standpipe.

Wind Chill Index   See Chill Factor.

Wye Connection   See Star Connection.

Zeolite   A substance used in water softening, in which a filtering tank is recharged by passing a salt solution through it. The tank may then be used as a filter for the free ions associated with hard water.

Zone   A portion of a building controlled by a single thermostat, because its spaces have similar heating or cooling needs.

BIBLIOGRAPHY

The following list of books is provided for candidates who may wish to do further research or study in Mechanical and Electrical Systems. Most of the books listed below are available in college or technical bookstores, and all would make welcome additions to any architectural bookshelf. In addition to the course material and the volumes listed below, we advise candidates to review regularly the many professional journals, which are available at most architectural offices.

ANSI A117.1
(Handicapped Standards)
American National Standards Institute

Architectural Graphic Standards
Ramsey and Sleeper
Wiley and Sons

ASHRAE Handbook of Fundamentals
American Society of Heating, Refrigerating and Air-Conditioning Engineers

Concepts in Architectural Acoustics
M. David Egan
McGraw-Hill

Design with Climate
Victor Olgyay
Princeton University Press

IES Handbook Reference Volume
John Kaufman
Illuminating Engineering Society

Mechanical and Electrical Equipment for Buildings
Stein/Reynolds/McGuinness
Wiley and Sons

National Electrical Code
National Fire Protection Association

National Plumbing Code
American Society of Mechanical Engineers

The Passive Solar Energy Book
Edward Mazria
Rodale Press

Uniform Building Code

Lesson 1

1. **D** All of the above. The holistic approach to sustainably designed projects encourages the design team to examine the impact of environmental, economic, mechanical, and aesthetic architectural decisions.

2. **D** None of the above.

   Choice I is not correct. The zone of the earth that supports human life (five miles into the earth's crust and five miles into the atmosphere) is an extremely fragile ecosystem. This biosphere that has evolved over millions of years has been dramatically affected by the growth of human activity in the last 150 years.

   Choice II is not correct either. While innovative technologies are improving energy efficiency of some building systems, the vast majority of the built environment is energy inefficient.

   Choice III is also not correct. Toxic substances have the tendency to expand and affect large areas. For example, the air above the Great Lakes contains evidence of DDT, a toxic pesticide banned in the United States decades ago. It was discovered that DDT is captured in the jet stream bringing toxic materials from far away continents, which still use toxic pesticides.

   Choice IV is not correct. While recycling is helpful, it is just the beginning of the sustainable design process. The principles of sustainable design say that we need to have more building products that can be recycled and are biodegradable to create a more sustainable ecosystem.

3. **C** I, III, and IV

   Choice I is correct. Designing with native landscaping is preferred to using exotic or imported plant types. Indigenous plants tend to survive longer, use less water, and cost less.

   Choice II is not correct. Placing any structure in a floodplain, even those that resist floodwater, is not desirable. Placing buildings in a floodplain can increase flooding farther down stream.

   Choice III is also correct. Buildings sensitive to the benefits of solar orientation and passive and active solar gain techniques save energy and are more visually aligned with local climatic conditions.

   Choice IV is correct as well. In-fill development and proximity to a variety of transportation options are design principles that benefit the inhabitants and their environment.

4. **C** III and IV

   I is not correct. Communities that are only residential are not encouraged. Mixed-use development (combining housing, retail, open space, and commercial) is a preferred sustainable design.

   II is not correct. Open space should not be designed only for recreation and wildlife habitat. Additional uses such as environmental education, storm water retention, flood control, wetlands drainage, and so on, should be considered in sustainable planning.

   III is correct. The Ahwahnee principles support a wide range of interconnected transportation to encourage many options for travel.

   IV is also correct. Development that permits opportunities for a diverse number of jobs is a key goal of the Ahwahnee Principles.

5. **D** All of the above

   I is correct. While first cost is not the primary concern of life cycle costing, it is one of the economic factors considered.

   II is also correct. The cost of maintenance is part of the evaluation.

III is correct as well. The durability of a product or system is considered in the cost of repair and part of the overall evaluation.

IV is correct because the comparison of product or system life is one of the factors evaluated in life cycle costing.

6. **B** I, II, and III

I is correct. LEED has several options for improving IAQ (Indoor Air Quality) including filtering the air system and installing low VOC (Volatile Organic Compound) paints and caulking.

II is also correct. Methods to store, recirculate, and locally distribute rainwater are encouraged.

III is correct as well. Innovative solutions to energy conservation such as fuel cells, photovoltaic panels, and gas turbine energy production are encouraged in the LEED accreditation system.

IV is incorrect. Unfortunately, the LEED system awards no points awards for designs with strong aesthetics.

7. **D** All of the above

All of these consultants (wetlands engineer, energy commissioner, landscape architect, and energy modeling engineer) might be necessary for the holistic approach to sustainable design. The landscape architect should have experience with local, native plant design.

8. **C** I, II, and IV

I is correct. Computer programs that allow energy modeling of design options allow the architect a quick method of evaluating numerous different solutions.

II is also correct. It is extremely important that the client be able to understand the value of sustainable design solutions.

III is not correct. Art selection is at the client's discretion.

IV is correct. Locating the most energy efficient appliances, plumbing fixtures, and office equipment will improve the energy efficiency of the entire project.

9. **B** I, III, and IV

I is correct. Solar orientation can affect many architectural design elements including massing, landscaping, fenestration, and building skin design.

II is not correct. Landscape design should be functional as well as visually pleasing. Landscape design for purely visual impact is not consistent with the sustainable design approach.

III is also correct. Architectural design that understands the context (scale, color, style, texture) of adjacent structures is sympathetic to the sustainable design philosophy.

IV is correct as well. Understanding all site conditions, and their potential to assist building's energy systems is helpful. For example, ground water connected to a heat pump is a good source of supplemental energy for cooling and heating a building.

10. **A** I, II, and IV

I is correct. Solar shading, whether from landscaping or architectural elements, can regulate the insulation to increase winter light and reduce warm summer sunlight.

II is also correct. Urban heat island effect is the tendency of a building roof to absorb solar radiation during the day and then emit heat radiation during the evening. Roof systems with grass or light colored roofing material reduce the urban heat island effect.

III is not correct. Sustainable design encourages approaches that reduce the area allocated to parking.

IV is correct. The type, location, and size of building fenestration are key aspects of architectural design for sustainable projects.

Lesson 2

1. **C** Rain water is slightly acidic and tends to corrode metal pipes. Furthermore, contaminants may be absorbed into the rain water, which may make it unhealthful, while ground water tends to be partially filtered of such contaminants.

2. **C** When water seeps into the ground, it dissolves minerals and becomes hard. This is not usually hazardous to humans, but tends to clog pipes. Hardness of water is not the same as acidity; they are different problems and are treated differently.

3. **D** Acidic water causes corrosion, not deposition. Where water is acidic, corrosion-resistant materials such as copper and PVC are appropriate for piping. Hardness, not acidity, can be controlled by the zeolite process. Plastic pipe, including PVC, should never be used in exposed locations above ground because it deteriorates when exposed to ultra-violet light, which is present in direct sunlight.

4. **A** The lower the pH reading, the more acidic the water. 5 is very low for water. A pH of 7 is neutral. There is no such thing as pH 17.

5. **B** Copper is soldered, which relies on heat to melt the flux and solder.

6. **D** 100 psi − 10 psi = 90 psi at the building. The faucet requires 10 psi, leaving 90 − 10 = 80 psi to lift the water within the building. 80 psi × 2.3 ft/psi = 184 feet.

7. **B** II only. I is true of both the downfeed and the pneumatic systems. II is true only of the downfeed. III is not true.

8. **B** II only. I is not true. II is true, and is the reason for separation. III is not true; in general, neither storm drainage nor sanitary drainage is pressurized.

9. **D** The water in the trap keeps the methane gas that is generated in the sewers below from rising up into the building.

10. **A** A catch basin is an inlet at the low point of a swale or other drainage collection area to collect surface runoff. (An interceptor collects grease, etc.)

Lesson 3

1. **A** Insolation (solar gain) is a form of radiation. It does not require contact (the earth does not touch the sun), is transmitted through clear objects such as glass, and can be blocked by opaque objects.

2. **C** Convection is the only heat transfer process in which *direction* (up, down, or sideways) makes any difference, and therefore is the only process in which relative height would make a difference.

3. **B** The MRT is defined as the weighted average of the surrounding surface temperatures.

4. **A** The CLTD is based on wall mass, orientation, and color, which relate to conduction and radiation.

5. **D** The resistance determines the temperature gradient and the conduction, and is determined by the conductivity.

6. **C** A degree day measures the temperature over a period of time against a reference temperature (65°F). A design day is a theoretical day that is "worse" (either colder or hotter) than 98 percent of the days experienced at a site, and allows us to size the plant to cover most days.

7. **D** See the answer to the above question and page 53.

8. **C** A and B are immediately eliminated because q_{CLTD} is always a heat gain. D

contains q$_r$, which is a gain from the sun. C contains *only* heat *losses*.

9. **A** Design day temperatures can only refer to a particular moment in time, eliminating B and D. Temperature does not measure latent heat, eliminating C. Design days are expressly used for equipment sizing.

10. **D** The evaporation of steam is a prime example of latent heat transfer. In addition, heat is conducted into the pot from whatever it is sitting on, and the heat is drawn away by the convective flow in the air around the pot (and under the pot, if it is a gas stove).

Lesson 4

1. **A** The ambient air temperature must make up for the radiant loss to the surroundings.

2. **D** The *relative* humidity changes with temperature. All of the other choices are absolute measures of the amount of water in the air, and say nothing about how close the air is to saturation.

3. **C** A sling psychrometer is an instrument designed to measure the wet bulb temperature. This is compared with the dry bulb temperature to determine the relative humidity.

4. **C** There are several kinds of thermal storage walls, but a Trombe wall is distinguished by the fact that it causes a convective loop.

5. **D** Hot arid climates often experience a large diurnal temperature swing, so damping or averaging temperature swings is an excellent strategy. If there is a water supply, there will be a cooling effect from evaporation. Unfortunately, if the outside air is hot, there is little benefit from ventilation. This is especially true if the other two strategies succeed in cooling, since the

inside air will then be cooler than the outside air. (Note: Hot *humid* climates are more suited for ventilation, where that is often the only passive possibility.)

6. **C** The declination angle is affected only by the time of year, not the time of day.

7. **A** Refer to the psychrometric chart. Find where the 60°F vertical line crosses the 70 percent RH curved line and follow the diagonal line upward and to the left to read 54°F.

8. **A** There is no change in enthalpy due solely to air movement.

9. **A** The earth stores heat lost into it, and changes temperature much more slowly than outside air. However, earth is not a good insulator, moisture usually causes problems, and the weight of the earth increases structural costs.

10. **D** Since focusing collectors must track the sun, they are usually more difficult to mount than flat plate collectors.

Lesson 5

1. **C** The four-pipe system can bring the heating and cooling capacity to the coils in the fan coil system.

2. **B** Systems that use the compressive refrigeration cycle reverse the flow from hotter to cooler objects (C). A heat pump uses the refrigeration cycle (A). An evaporative chiller, such as a swamp cooler, can do the same (D), but is most often used for increasing the flow rate of heat away from the condenser coils on a compressive refrigeration cycle. The only incorrect choice is B.

3. **D** The electric reheat system (I) normally cools, but can heat when the resistance coil is on. The double duct system (II) distributes

only air and can heat or cool, as needed, at each mixing box. The fan coil system (III) can also heat or cool at each unit.

4. **A** The evaporative chiller uses the latent heat of evaporation to transfer heat to the outside air.

5. **A** Both the unitary system and the heat pump system can have a meter on each unit, thus keeping track of individual usage and permitting separate billing.

6. **C** The heat pump system uses a heat sink connected to several heat pumps. When one zone is cooling at the same time as another is heating, for example in the morning if one zone is in the easterly part of a building and another is in the west, then the system can balance itself without a boiler or chiller. This can happen fairly often in mild climates.

7. **C** The condenser is the high pressure coil, which is hottest and gives off the most heat.

8. **B** The one pipe system is run in series, while the others are parallel systems.

9. **C** The one and two pipe systems provide either heating or cooling, but not both simultaneously. The three and four pipe systems can provide both simultaneously.

10. **A** It is called return air because it is *returning* to the *plant*.

Lesson 6

1. **A** $I = V/R = > 120 \text{ v}/8 \ \Omega = 15$ amps

2. **A** $P = V \times I \times PF = > 120 \times 15 \times 1 = 1,800$ watts

3. **D** $R = 1/(1R_1 + 1/R_2 + 1/R_3)$
$= 1/(1/4 + 1/4 + 1/2)$
$= 1/1$
$= 1.0$ ohms

4. **B** $V_2 = V_1/\sqrt{3}$
$V_1 = V_2 \times \sqrt{3} = 120 \text{ v} \times 1.73 = 208 \text{ v}$

5. **C** There are three wires from the secondary of a three wire transformer: a line, the neutral, and the opposite line. The secondary winding has two options for connection: one line to the other line, or either line to neutral.

6. **B** In this case, answer B, three-way switches and four-way switches, is correct. When more than two switches are necessary, two of the switches must be three-way switches and the remaining additional switches must be four-way switches.

7. **A** Universal motors are the typical motors in small appliances, allowing variable speeds. See page 121.

8. **D** Armored cable may not be embedded in concrete.

9. **D** A ground fault interrupter is specifically designed to pick up low level electrical short circuits, which often occur with no equipment switched "on," and then shut the entire circuit off.

10. **C** A *demand* surcharge is based on the peak usage, or demand, and encourages the customer to reduce those peaks. Reduced peaks reduce the need for new power generating plants.

Lesson 7

1. **D** The "warm" colors come from the *lower* Kelvin temperatures. Psychologically warm colors happen to be generated by lower temperatures than the psychologically cool colors.

2. **C** The retina is the sensing surface. The lens focuses, the iris adjusts the amount of light, and there are no cilia in the eye (there is a ciliary muscle).

3. **B** Glare comes from high contrast. It can occur at low light levels, for example from headlight glare at night. It can come directly from a source, not just from reflections, for example by looking directly at the sun.

4. **B** Translucent means that light can pass through, but not an image.

5. **C** Ambient lighting is usually diffuse, and diffuse lighting does not cause veiling reflections, so it is a good light for reading flat surfaces. Mirrors are not diffusing, but specular. A flat paint would be a diffuse reflecting surface.

6. **C** A footcandle is a measure of illumination. Candlepower measures the power at the source, a lumen measures flux, and a footLambert measures reflected or transmitted brightness.

7. **A** According to the inverse square law, the illumination at 20 feet is $(10/20)^2$ that at 10 feet, or 1/4 of 24 fc, or 6 fc.

8. **C** Low voltage lamps have smaller filaments, and therefore allow more precise optical control of light placement.

9. **D** An R-40 bulb is $40 \times 1/8"$ or 40/8 or 5" in diameter.

10. **D** All three statements are true.

Lesson 8

1. **C** Noise criteria or NC curves, see page 156.

2. **A** IL or intensity level, see page 147.

3. **B** The sound transmission class (STC) is a widely accepted method of rating walls and doors according to their typical or

overall resistance to sound transmission. See page 181.

4. **A** The reverberation time is defined as the time required for a sound to decay 60 dB in a space. The return time for an echo depends on the distance to the reflecting surface, and is only vaguely related to reverberation time. Reverberation time is *shorter* in a dead space.

5. **C** 40 dB is acceptable for a hospital or a quiet office. 10 dB is not a practical level, and 70 dB is definitely too loud.

6. **A** Doubling the distance results in a decrease of 6 dB.

7. **D** Doubling the source power results in an increase of 3 dB.

8. **A** Doubling the mass theoretically results in a 6 dB drop.

9. **D** Carpeting and drapes are much more absorptive surfaces than concrete. Increasing the absorptivity decreases both the reverberation time and the volume.

10. **C** The STC is comparatively high, so the wall is good at stopping transmitted sound. The STC says nothing about absorptivity or reflected sound.

Lesson 9

1. **D** The maximum usable speed of an electric elevator varies from about 300 to 1,800 feet per minute, and is limited by the building height, as acceleration and deceleration require a substantial distance.

2. **D** Buildings less than 50 feet high, or five stories high, generally have hydraulic, rather than electric, elevators. Hydraulic elevators require no penthouse and are simpler and less expensive to install.

3. **C** Refer to the table on page 173.

4. **A** Fire stairs required for emergency exiting must have fire doors that open in the direction of emergency travel.

5. **B** Freight elevators are generally slower and have greater capacity than passenger elevators. They may also have different types of doors. Freight elevators often carry people as well, and must have the same safety features as passenger elevators.

6. **C** The maximum slope permitted for a handicapped ramp is 1:12. Thus, if the difference of elevation is 2 feet, the length of the ramp must be at least 12×2 feet, or 24 feet.

7. **B** The main brake is mounted on the motor shaft of the elevator machine and stops the car automatically in case of a power failure. The governor measures and limits the elevator speed. The interlocks prevent the lobby door from opening when the elevator car is elsewhere. And the car bumpers, located at the bottom of the shaft, stop the car if it should over-travel at low speed.

8. **A** Moving ramps allow pedestrians to move swiftly: if one continues to walk along the moving ramp, one's speed will be doubled.

9. **C** The number of elevators required in a new building is determined by the number of floors (I), volume of traffic (II), and size and capacity of the elevator cars (V). Elevator traffic is computed for peak traffic periods and is not based on the number of hours of operation (III is incorrect), and the elevator roping system is not related to the number of elevators required (IV is incorrect).

10. **D** Although all four answers have some degree of truth, escalators are most effectively used where large numbers of people must be moved quickly.

Lesson 10

1. **B** Compartmentation attempts to contain the fire within the area in which it started for a given time, so that occupants may safely escape the building, damage spread may be limited, and the fire department may combat the blaze.

2. **C** 44 inches is the *minimum* exit corridor width.

3. **C** The maximum distance from a room to an exit stairway is 150 feet unless the building is sprinklered. See page 184.

4. **B** The hose is 100 feet long, and every point on the floor must be reached by a 30-foot spray from that hose.

5. **A** Hot air and smoke from small or smoldering fires often heat the ceiling evenly but slowly. When the temperature finally goes above ignition, it does so for nearly the entire ceiling at once. This is called flashover.

6. **C** Occupancy groups are defined by function. Construction materials determine the building construction type.

7. **C** The fire department must pump water into a dry standpipe to use it for fighting fires. Building occupants have no way to use them, unlike wet standpipes. Halon gas or carbon dioxide may fill an extinguisher system similar to a sprinkler system, but not a standpipe. There is no special drainage system provided for fighting fires.

8. **A** Sprinklers function whether or not someone is there to observe the fire. They are more expensive than the standpipes, they can damage artwork, books or records, and they may *not* be used as part of the domestic plumbing.

9. A The ionization detector will detect invisible products of combustion. Smoke detectors only detect smoke of significant density. A fusible link only melts when it gets hot enough, which may be after some other part of the space is already well involved in the fire. The same is true of a heat detector.

10. D All three are required to have a siamese connection at the base, so that the fire department can add water to the system.

The examination on the following pages should be taken when you have completed your study of all the lessons in this course. It is designed to simulate the Mechanical and Electrical Systems division of the Architect Registration Examination. Many questions are intentionally difficult in order to reflect the pattern of questions you may expect to encounter on the actual examination.

You will also notice that the subject matter for several questions has not been covered in the course material. This situation is inevitable and, thus, should provide you with practice in making an educated guess. Other questions may appear ambiguous, trivial, or simply unfair. This too, unfortunately, reflects the actual experience of the exam and should prepare you for the worst you may encounter.

Answers and complete explanations will be found on the pages following the examination, to permit self-grading. **Do not look at these answers until you have completed the entire exam.** Once the examination is completed and graded, your weaknesses will be revealed, and you are urged to do further study in those areas.

Please observe the following directions:

1. The examination is closed book; please do not use any reference material.

2. Allow about 60 minutes to answer all questions. Time is definitely a factor to be seriously considered.

3. Read all questions *carefully* and mark the appropriate answer on the answer sheet provided.

4. Answer all questions, even if you must guess. Do not leave any questions unanswered.

5. If time allows, review your answers, but do not arbitrarily change any answer.

6. Turn to the answers only after you have completed the entire examination.

GOOD LUCK!

EXAMINATION ANSWER SHEET

Directions: Read each question and its lettered answers. When you have decided which answer is correct, blacken the corresponding space on this sheet. After completing the exam, you may grade yourself; complete answers and explanations will be found on the pages following the examination.

1. Ⓐ Ⓑ Ⓒ Ⓓ
2. Ⓐ Ⓑ Ⓒ Ⓓ
3. Ⓐ Ⓑ Ⓒ Ⓓ
4. Ⓐ Ⓑ Ⓒ Ⓓ
5. Ⓐ Ⓑ Ⓒ Ⓓ
6. Ⓐ Ⓑ Ⓒ Ⓓ
7. Ⓐ Ⓑ Ⓒ Ⓓ
8. Ⓐ Ⓑ Ⓒ Ⓓ
9. Ⓐ Ⓑ Ⓒ Ⓓ
10. Ⓐ Ⓑ Ⓒ Ⓓ
11. Ⓐ Ⓑ Ⓒ Ⓓ
12. Ⓐ Ⓑ Ⓒ Ⓓ
13. Ⓐ Ⓑ Ⓒ Ⓓ
14. Ⓐ Ⓑ Ⓒ Ⓓ
15. Ⓐ Ⓑ Ⓒ Ⓓ
16. Ⓐ Ⓑ Ⓒ Ⓓ
17. Ⓐ Ⓑ Ⓒ Ⓓ
18. Ⓐ Ⓑ Ⓒ Ⓓ
19. Ⓐ Ⓑ Ⓒ Ⓓ
20. Ⓐ Ⓑ Ⓒ Ⓓ
21. Ⓐ Ⓑ Ⓒ Ⓓ
22. Ⓐ Ⓑ Ⓒ Ⓓ
23. Ⓐ Ⓑ Ⓒ Ⓓ
24. Ⓐ Ⓑ Ⓒ Ⓓ
25. Ⓐ Ⓑ Ⓒ Ⓓ
26. Ⓐ Ⓑ Ⓒ Ⓓ

27. Ⓐ Ⓑ Ⓒ Ⓓ
28. Ⓐ Ⓑ Ⓒ Ⓓ
29. Ⓐ Ⓑ Ⓒ Ⓓ
30. Ⓐ Ⓑ Ⓒ Ⓓ
31. Ⓐ Ⓑ Ⓒ Ⓓ
32. Ⓐ Ⓑ Ⓒ Ⓓ
33. Ⓐ Ⓑ Ⓒ Ⓓ
34. Ⓐ Ⓑ Ⓒ Ⓓ
35. Ⓐ Ⓑ Ⓒ Ⓓ
36. Ⓐ Ⓑ Ⓒ Ⓓ
37. Ⓐ Ⓑ Ⓒ Ⓓ
38. Ⓐ Ⓑ Ⓒ Ⓓ
39. Ⓐ Ⓑ Ⓒ Ⓓ
40. Ⓐ Ⓑ Ⓒ Ⓓ
41. Ⓐ Ⓑ Ⓒ Ⓓ
42. Ⓐ Ⓑ Ⓒ Ⓓ
43. Ⓐ Ⓑ Ⓒ Ⓓ
44. Ⓐ Ⓑ Ⓒ Ⓓ
45. Ⓐ Ⓑ Ⓒ Ⓓ
46. Ⓐ Ⓑ Ⓒ Ⓓ
47. Ⓐ Ⓑ Ⓒ Ⓓ
48. Ⓐ Ⓑ Ⓒ Ⓓ
49. Ⓐ Ⓑ Ⓒ Ⓓ
50. Ⓐ Ⓑ Ⓒ Ⓓ
51. Ⓐ Ⓑ Ⓒ Ⓓ
52. Ⓐ Ⓑ Ⓒ Ⓓ
53. Ⓐ Ⓑ Ⓒ Ⓓ

1. Which of the following does an open fireplace contribute to the environment?

   I. Radiant heat

   II. Convection

   III. Latent heat

   A. I only   **C.** III only

   B. II only   **D.** I and II

2. Given a coefficient of utilization for a given room of .54, a maintenance factor of .8, and a typical four lamp fixture with 2,900 lumens/lamp, how many fixtures are needed to provide a light level of 50 footcandles if the room size is 100×100 feet?

   A. 10 fixtures   **C.** 60 fixtures

   B. 30 fixtures   **D.** 100 fixtures

3. Which of the following strategies reduces overall energy consumption in a building?

   I. An HVAC system with an economizer cycle

   II. A daylighting design which provides light but little heat

   III. A wet standpipe system

   A. I and II   **C.** I and III

   B. II and III   **D.** I, II, and III

4. Which of the following plumbing devices most affects the safety and health of a building's occupants?

   A. A gate valve

   B. A globe valve

   C. A trap

   D. An interceptor

5. Which of the following is true?

   A. When DB equals WB, RH equals 100 percent.

   B. DB cannot equal WB.

   C. DB cannot be greater than WB.

   D. RH can exceed 100 percent.

6. Which of the following is NOT usually true of a passive solar design?

   A. It blocks the sun in overheated months.

   B. It stores heat in thermal mass or phase change materials.

   C. It is cheaper to build than a conventional building.

   D. It relies on a backup system for dealing with extremes in weather.

7. Which of the following statements is NOT true of a heat pump?

   A. It works in the same manner as an air conditioner.

   B. It cannot deliver more heat energy than the energy required to run the pump.

   C. It works only between certain temperature ranges.

   D. It is often used in conjunction with a recirculating heat sink.

8. What is the primary purpose of a leach field?

   A. To store storm runoff until it can seep into the ground

   B. To store sanitary drainage until it can be pumped out and removed

   C. To remove greywater before it enters the public sewage system

   D. To allow sewage cleared of solid matter to seep into the ground

9. Reverberation time depends on which of the following factors?

   I.   Room volume

   II. Sound volume

   III. Absorptivity of surfaces

   A. I and II   **C.** I and III

   B. II and III   **D.** I, II, and III

10. One of the following is an appropriate lighting design. Which is it?

    A. A fluorescent cove light, casting indirect light on the ceiling, is used to bring out the sparkle in a display of diamonds.

    B. A mercury vapor lamp is used in an operating room to get the highest possible light levels and proper color rendition of the patient.

    C. Suspended incandescent globes are used in a computer room to reduce the sterile atmosphere and provide warm colors that do not reflect in the video screens.

    D. Low voltage lamps are used in lighting a display because of their tight focus.

11. Which of the following are included in the refrigeration cycle?

    I.   Freon

    II. Compressor

    III. Expansion valve

    A. I and II   **C.** I and III

    B. II and III   **D.** I, II, and III

12. Rank the following walls in the order of their ability to stop the transmission of sound.

    I.   A concrete wall

    II. A stud wall with plaster on both faces

    III. A stud wall with plaster and an extra layer of acoustical tile on both faces

    A. III, II, I   **C.** II, III, I

    B. I, III, II   **D.** III, I, II

13. Rank the following materials in the order of their ability to absorb sound.

    I.   Carpet

    II. Plaster

    III. Acoustical tile

    A. III, II, I   **C.** II, III, I

    B. I, III, II   **D.** III, I, II

14. Which of the following is UNRELATED to the definition of degree day?

    A. History of outside temperature

    B. Design day temperature

    C. Reference temperature

    D. Number of hours per day at a given temperature

15. What is the purpose of the trap in a sewage system?

    A. To keep sewer gas from passing up into the building

    B. To catch small valuables and other materials before they pass down the drain

    C. To catch grease before it clogs the drain

    D. To provide a cleanout when drains become clogged

16. Which of the following is NOT one of the purposes of a vent stack?

    A. To provide a vent for sewer gases

    B. To keep the water in the trap from getting siphoned out

    C. To vent several different drains

    D. To provide a vent for the bathroom and its odors

17. Which of the following is NOT a reasonable strategy for a building located in a hot arid climate?

    A. Provide maximum ventilation to the outside to dissipate internal heat gain.

    B. Provide high thermal mass walls.

    C. Provide an internal courtyard with shading devices and/or evaporation sources.

    D. Place homes in close proximity for mutual shading.

18. Which of the following are NOT fire extinguishing agents?

    I.   Halon

    II. Freon

    III. Carbon Dioxide (CO_2)

    IV. Metal Halide

    A. I and II   C. II and III

    B. III and IV   D. II and IV

19. Given a 20 × 30 × 10 foot room, with concrete floor, walls and ceiling, what is the total absorptivity of the room, if the absorption coefficient for concrete is 0.02 sabins?

    A. 24 sabins   C. 84 sabins

    B. 44 sabins   D. 120 sabins

20. Which of the following is NOT a high intensity discharge lamp?

    A. Tungsten halogen

    B. High pressure sodium

    C. Mercury vapor

    D. Metal halide

21. If one Evinrude outboard motor results in 70 dB of sound in the boat, what is the sound level of two motors?

    A. 64 dB   C. 76 dB

    B. 73 dB   D. 140 dB

22. If an outboard motor results in a sound intensity level of 60 dB at a distance of 40 feet, what is the IL at 80 feet?

    A. 30 dB   C. 54 dB

    B. 40 dB   D. 57 dB

23. A proposed building has a peak cooling load of 144,000 Btuh for the design day. How many tons of air conditioning does this represent?

    A. 12 tons   C. 144 tons

    C. 72 tons   D. 288 tons

24. What is the function of a catch basin?

    A. It holds overflow until it can be drained away later.

    B. It collects the water under a fixture (it is often called a sink).

    C. It collects surface runoff and admits it to a storm drainage line.

    D. It intercepts grease and solid objects before they clog a main sewer line.

25. If a wall has a total thermal resistance of 21.6 s.f. °F/Btuh (including air films), what is the U value of the wall?

   A. 0.046 Btuh / ft^2 °F

   B. 0.46 Btuh / ft^2 °F

   C. 2.16 Btuh / ft^2 °F

   D. 21.6 Btuh / ft^2 °F

26. What is the likely result of excessively hard water in piping?

   A. Steel pipe will corrode.

   B. Plastic pipe will fill with encrustations.

   C. Copper pipe will rust.

   D. Steel pipe will plug up with depositions.

27. What is the likely result of excessively acidic water in piping?

   A. Steel pipe will corrode.

   B. Plastic pipe will fill with encrustations.

   C. Copper pipe will rust.

   D. Steel pipe will plug up with depositions.

28. What is the special danger of a class C fire?

   A. There will be a chemical reaction between the burning objects and the extinguisher medium.

   B. The user of the extinguisher may be electrocuted.

   C. The fire will spread more rapidly than any other fire.

   D. There is no special danger, since class C is the least hazardous type of fire.

29. Which of the following statements is correct?

   A. A dry pipe sprinkler system causes less water damage when discharging and is therefore preferable to a wet pipe system.

   B. A POC detector responds only to invisible gases.

   C. A fire alarm should be both audible and visible.

   D. A dry standpipe should be connected to a dependable water source at all times.

30. In remodeling an existing school building that has a steam heat system and no air conditioning, the intent is to create separate living units. Which of the following systems is probably the best choice for the building?

   A. A double duct forced air system, which would provide each unit either heating or cooling, independent of the other units.

   B. A variable air volume system, which would be the most efficient, even if one unit is being heated while an adjacent unit is being cooled.

   C. A heat pump system, which would not require the addition of ductwork, and would allow each unit to be heated or cooled, as desired.

   D. A multizone system, which would provide for different temperature zones with less space required than any of the others.

31. Which of the following systems includes an evaporator coil and a condenser coil?

   I. Fan coil system

   II. Heat pump system

   III. Refrigeration cycle

   A. I and II   **C.** I and III

   B. II and III   **D.** I, II, and III

32. What system or equipment uses only evaporation and no compression to cool water?

   A. Double duct system

   B. Fan coil system

   C. Unitary system

   D. Evaporative chiller

33. Which of the following terms means brightness?

   A. Luminance   **C.** Glare

   B. Illumination   **D.** Radiation

34. In designing a hotel, which of the following must be provided in at least some of the bathrooms to make them accessible to the handicapped?

   I.   Grab bars

   II. Shower stall seat

   III. Shower stall large enough to accommodate a five-foot turning circle

   A. I and II   **C.** I and III

   B. II and III   **D.** I, II, and III

35. A single duct constant volume system with electric reheat has all the following characteristics, EXCEPT

   A. its initial cost is low.

   B. it can heat or cool, but not both simultaneously.

   C. it is inefficient if there is very much need for heating.

   D. it includes an electric resistance heater adjacent to the cooling coil.

36. A fitting outside a building which provides two or more connections through which water can be pumped to a standpipe or sprinkler system is called a

   A. hydrant.

   B. ball drip.

   C. siamese fitting.

   D. combination standpipe.

37. The intensity of a light source is measured in what units?

   A. Lumens   **C.** Footcandles

   B. Candlepower **D.** Lamberts

38. What system removes heat from the air exhausted from a building and transfers it to the incoming fresh air?

   A. Fan coil system

   B. Heat pump system

   C. Air-to-air heat exchanger

   D. Variable air volume system

39. What is the thermal process which takes place in a fluid medium?

   A. Radiation   **C.** Convection

   B. Conduction   **D.** Transmission

40. The ability of a surface to emit heat by radiation relative to that of a black body at the same temperature is called

    A. radiation.
    B. emissivity.
    C. transmissivity.
    D. reflectivity.

41. Which system always has both heated air and chilled air running through it?

    A. Double duct system
    B. Multizone system
    C. Fan coil system
    D. Variable air volume system

42. A stack vent is

    A. the same as a vent stack.
    B. the air intake line for all the fixtures.
    C. a pipe into which all the soil and waste lines empty.
    D. the extension of a soil stack to the outdoors above the highest branch drain.

43. What factor affects both reverberation time and loudness in enclosed spaces?

    A. Acoustical intensity
    B. Transmissivity
    C. Absorptivity
    D. Frequency

44. Insolation is

    A. a material having high resistance to heat flow.
    B. heat gain from the sun.
    C. heat energy transmitted by radiation.
    D. thermal resistance.

45. Pitch as perceived by the ear is determined by a sound's

    A. frequency.
    B. velocity.
    C. intensity.
    D. pressure.

46. CLTD (cooling load temperature differential) includes the effects of all of the following, EXCEPT

    A. time lag.
    B. radiation.
    C. conduction.
    D. humidity.

47. Both sound intensity level (IL) and sound pressure level (SPL) are measured in

    A. watts.
    B. decibels.
    C. sones.
    D. hertz.

48. Which of the following systems can be used if each zone is to be billed separately?

    I.   Heat pump system
    II. Multizone system
    III. Unitary system

    A. I and II
    B. II and III
    C. I and III
    D. I, II, and III

49. A step down transformer in a building is used to

    A. decrease the voltage.
    B. decrease the current.
    C. decrease the power.
    D. convert alternating current to direct current.

50. The ratio of the actual amount of water vapor in the air to the maximum amount that the air could contain at the same temperature is called the

    A. humidity ratio.
    B. effective humidity.
    C. enthalpy.
    D. relative humidity.

51. The *Natural Step* is an approach to the environment that follows which of the following principles?

   I.   The biosphere affecting humans is a relatively stable and resilient zone which includes five miles into the earth's crust and five miles into the troposphere.

   II. Improved technologies have dramatically increased the number and quantity of available natural resources.

   III. Toxic substances released into either the sea or atmosphere will only influence areas adjacent to the toxic source.

   IV. Using building materials that are recycled is an adequate sustainable design approach.

   A. I

   B. II

   C. II and IV

   D. None of the above

52. Which of the following is a consultant that might be employed in a sustainable design project?

   I.   Wetlands engineer

   II. Energy commissioner

   III. Landscape architect

   IV. Energy modeling engineer

   A. I

   B. I and II

   C. I, III, and IV

   D. All of the above

53. Sustainably designed architecture requires attention to which of the following building elements?

   I.   Solar shading devices

   II. Urban heat island effect

   III. Increased parking

   IV. Fenestration and glazing

   A. I, II, and IV

   B. I and IV

   C. I and II

   D. All of the above

54. Which of the following features are required in an elevator car in order for the car to comply with the ANSI handicapped standards?

   I.   Self-leveling device

   II. Door reopening device

   III. Emergency communication system

   IV. 36-inch-high car control buttons

   V. Hard, non-slip floor surface

   A. I and II   **C.** II, IV, and V

   B. I, III, and V   **D.** I, II, III, and IV

55. Among the following factors, which are considered distinct advantages in the use of escalators?

   I.   They operate continuously.

   II. They have low power consumption.

   III. They are dependable in the event of fire.

   IV. They are architecturally flexible.

   V. They are relatively quiet.

   A. I, II, and IV   **C.** I, II, and III

   B. I, II, and V   **D.** III, IV, and V

56. What is the desirable interval for elevators in a high-rise office building?

   A. 5 seconds   **C.** 2 minutes

   B. 30 seconds   **D.** 5 minutes

The examination answers and explanations will be found on the following pages.

Do not look at the answers until you have completed the exam.

EXAMINATION ANSWERS

1. **D** There is radiant heat from the hot gases and the embers (I), and the gases themselves rise rapidly (convection). Even the warm fireplace causes the air next to it to rise (II). However, little or no evaporation takes place outside of the combustion itself (III is incorrect).

2. **D** The formula for the illumination in a room is:

   E = (fixtures × lamps/fixture × lumens/lamp × maintenance factor × CU)/area

   Solving for the number of fixtures, given the desired illumination:

   Fixtures = (area × E)/(lamps/fixture × lumens/lamp × maintenance factor × CU)

   Fixtures = (10,000 s.f. × 50 fc) / (4 lamps/fixture × 2,900 lumens/lamp × .8 × .54)

   = 99.77 fixtures, which we round up to 100 fixtures

3. **A** An economizer cycle shuts off the refrigeration and uses outside air when the temperatures are low enough, thus reducing the plant load (I). Daylighting reduces the need for artificial lighting, which reduces the energy required. However, daylighting may also increase the heat gain and thus the cooling load, but that is not the case here, since we are told that the design provides little heat (II). A standpipe system is a fire safety device that does nothing until there is a fire, and thus has little to do with energy consumption (III).

4. **C** The trap is the device that keeps the sewer gas from entering the building, which would be harmful to the occupants. The gate and globe valves (A and B) simply regulate water flow. The interceptor (D) keeps solid objects and grease out of the sewer system, which affects the sewage treatment plant, but not the occupants.

5. **A** When the dry bulb temperature DB (not the decibel level, which is dB) is equal to the wet bulb temperature WB, then there is no evaporation cooling the sock on the psychrometer. This means that the relative humidity RH is 100 percent (A is correct, B is incorrect). When the RH is less than 100 percent, DB is greater than WB (C is incorrect). Finally, the relative humidity can never be greater than 100 percent, by definition (D is also incorrect).

6. **C** While passive solar designs often have lower operating costs, their first cost is generally greater, because of the required thermal mass and insulation. In passive solar design, the sun is blocked in overheated months (A), and thermal mass or phase change materials are used to store solar energy without large temperature swings (B). However, a backup system is still necessary (D), because extreme storm conditions are almost certain to occur sometime during the life of the building.

7. **B** A heat pump *does* deliver more energy than the energy required to run it. That is the whole point of the system. It actually moves heat from one point to another, rather than creating heat. It operates on the same cycle as an air conditioner (A), only with the inside and outside reversed. If the temperature range goes beyond the rated range of the refrigerant, the system won't work (C). Several heat pumps will often operate in conjunction with the same heat sink, recirculated within a large building or complex (D).

8. **D** A leach field lets the liquid sewage from a building seep into the ground. Surface runoff should be handled separately, since it would overload the system and back it up (A). A septic tank may need occasional emptying, but a leach field should not store anything (B). Greywater (waste from fixtures other than toilets and urinals) is typically recycled,

by sedimentation and filtering, and hence does not enter the leach field at all (C).

9. **C** Reverberation time is defined as the time it takes a 60 dB sound to die out after the source has stopped. It depends on room volume and surface absorptivity (I and III). It may vary with pitch, but only because absorptivity varies with pitch. However, it does not vary with sound volume (II).

10. **D** Low voltage lamps have smaller filaments, and therefore have more tightly controlled optics. This allows a more precise aim, which is useful in highlighting specific objects. A fluorescent cove light provides diffuse ambient illumination, ideal for reading or lighting flat surfaces evenly, but deadly dull in terms of sparkle (A). A mercury vapor lamp provides terrible color renditions of skin and flesh, even with very high light levels. This would be inappropriate or even dangerous in an operating room (B). Suspended incandescent globes provide warm colors and reduce the sterile atmosphere in a computer room, but they also cause glare spots all over the screens, making work very difficult (C). The fluorescent cove light at a low illumination level in choice A would be appropriate for the computer room.

11. **D** All three are included in the refrigeration cycle. See the diagram on page 94.

12. **B** Mass is the most important factor in stopping the transmission of sound, and the concrete wall has the greatest mass. Comparing the two stud walls, the acoustical tile will reduce *reflection*, but have minimal effect on *transmission*. However, since there is a slight increase in the amount of material present, the wall with the tile will represent a small improvement. The greatest benefit of the acoustical tile will be in the reduction of the sound level in the room containing the source, and the

reduction of reflected sound in the receiving room.

13. **D** The ability of a material to absorb sound is measured by its noise reduction coefficient (NRC). Acoustical tile has a NRC which ranges from .65 to .95, carpet ranges from .3 to .55, and plaster ranges from .02 to .05.

14. **B** A degree day is defined as the amount by which the average outdoor temperature at a particular location is below the reference temperature of 65°F for one day. Degree days may be summed and stated for a month or a year. Thus, A, C, and D all relate to the definition of degree day. The design day temperature (B) is a temperature which is lower than the outdoor temperature experienced at a given location on 98 percent of the days. It is therefore not directly related to the concept of degree day.

15. **A** The purpose of the trap is to keep methane gas generated by decomposition from entering the building. It is true that the trap may catch small valuables before they pass down the drain, but this is not the purpose of the trap. The trap catches grease, but this must be removed or dissolved, in order not to clog the trap. The trap may be removed when the drain is clogged, but this is a messy process, and the trap does not usually have a cleanout plug.

16. **D** The vent stack does not vent the bathroom itself, nor the odors that may occur there. It does vent sewer gases (A), and keep the water from getting siphoned out of the traps for several different drains (B and C).

17. **A** The outside air in a hot arid climate is too hot and dry during the day, and good design therefore limits air exchange with the outside. High thermal mass limits temperature swings, storing the heat gained during the day and reradiating it at night (B). A shaded courtyard, especially with some source of

evaporative cooling, provides a welcome relief to outside conditions (C). Mutual building shading is often beneficial (D), and many North African cities consist of a collection of contiguous mud brick buildings, with as little sun exposure as possible.

18. D   Freon (II) is a refrigerant, and metal halide (IV) is a type of high intensity discharge lamp. Carbon dioxide and halon are both used to smother fires without damaging the building's contents.

19. B   The absorptivity of the room is the surface area (not the volume) multiplied by the absorptivity of the surfaces.

$$A = \text{area} \times \alpha = [(2 \times (30 \times 20)) + (2 \times (30 \times 10)) + (2 \times (20 \times 10))] \times 0.02$$
$$= 44 \text{ sabins}$$

20. A   Tungsten halogen is an incandescent lamp.

21. B   Doubling the volume causes an increase of 3 dB. 70 dB + 3 dB = 73 dB.

22. C   Doubling the distance between source and receiver is equivalent to reducing the sound by one quarter, which results in a drop of 6 dB. 60 dB − 6 dB = 54 dB.

23. A   One ton is 12,000 Btuh. Therefore, 144,000/12,000 = 12 tons.

24. C   A catch basin admits water from a swale to a storm drainage line.

25. A   The U value is the reciprocal of the sum of the resistances. 1/21.6 = .046 Btuh/ft^2 °F.

26. D   Hard water means that there are minerals, typically calcium, dissolved in the water. These tend to deposit out in steel piping, clogging it. Corrosion is caused by acidic water (A), plastic pipe is not as subject to deposition as steel pipe (B), and copper pipe never rusts (C).

27. A   See the explanation for question 26.

28. B   A class C fire is an electrical fire, and there could be a short circuit that would conduct through water if water were used to put it out.

29. C   A fire alarm should be both audible and visible, so that blind or deaf people have some chance of perceiving it. A dry sprinkler system floods with water when it is set off (A). A POC detector (products of combustion) will respond to both visible and invisible gases (B). A dry standpipe is not connected to a water source until a fire occurs, when the fire department connects it to a hydrant via a pumper (D).

30. C   There are already pipes for the steam heat system. These may be used for the heat sink if they are in good condition, or they may be replaced by new pipes, which would be approximately the same size. All of the other systems are forced air systems that would require major modifications because the required ducting would occupy much more space. The VAV system (B) would require the least, but is not flexible enough to handle different units in different ways. To achieve that flexibility would require a double duct system, with two complete sets of ductwork (A), or a multizone system, with one separate branch for each unit (D). The heat pump system also allows each unit to be billed separately, based on the electricity used by that unit's heat pump.

31. B   The heat pump system and the refrigeration cycle both have an evaporator coil and a condenser coil.

32. D   The evaporative chiller, often called a cooling tower, uses only evaporation to cool water. Compression systems use evaporation and condensation in a closed loop.

33. A   Luminance means brightness. Illumination is the amount of light arriving at a surface, not leaving it. Glare results from extreme contrasts in brightness. Radiation is a method of transfer of heat or other energy.

34. A   Grab bars are required in all accessible shower stalls, and a seat is required in transfer type shower stalls. But neither a transfer type nor a roll-in type shower stall is required to be large enough to accommodate a five-foot turning circle.

35. D   All of the choices are characteristics of the single duct constant volume system with electric reheat, except D. An electric resistance heater is placed in the duct upstream of each diffuser, not adjacent to the cooling coil, to reheat the cooled air if heating is necessary.

36. C   A siamese fitting is one located close to the ground outside a building that provides two or more connections to permit the fire department to pump water to a sprinkler system or standpipe.

37. B   The intensity of a light source is measured in candlepower (B). The rate of flow of light is measured in lumens (A). A footcandle is a unit of illumination equal to one lumen per square foot (C). A lambert (D) is a unit of luminance or brightness.

38. C   The air-to-air heat exchanger removes heat from the exhaust air and warms the incoming fresh air, often at a very high efficiency.

39. C   Convection occurs in a fluid medium, often air or water.

40. B   Emissivity is the ability of a surface to emit heat by radiation relative to that of a black body at the same temperature. Black surfaces have higher emissivities than white or shiny surfaces.

41. A   A double duct system consists of two ducts, one carrying heated air and one carrying cooled air. The amount of air drawn from each duct at each room is controlled by dampers and mixed in a mixing box. The multizone system is similar, except that the mixing boxes are in the mechanical equipment room, and pre-mixed air at the desired temperature is sent out to each zone.

42. D   A stack vent is the extension of a soil stack to the outdoors above the highest branch drain. The air intake line for all the fixtures is called a vent stack, not a stack vent (A and B are incorrect). The pipe into which all the soil and waste lines empty is the soil stack (C is incorrect).

43. C   In acoustics, both the reverberation time and the loudness in an enclosed space are affected by the absorptivity of the surfaces in the space. Acoustical absorptivity is not the same as thermal absorptivity, nor are they measured in the same units.

44. B   Insolation refers to heat gain from the sun. Heat energy transmitted by radiation could be from the sun, or from some other source, such as a fireplace (C is incorrect). Both A and D refer to *insulation*, not insolation.

45. A   The frequency of a sound is the number of cycles per second of the sound wave, and is perceived by the ear as pitch.

46. D   CLTD is the model which uses the equation $q_c = U(A) \Delta T$, but substitutes CLTD for ΔT. CLTD adjusts for radiation and time lag in the basic conduction formula.

47. B   Both sound intensity level (IL) and sound pressure level (SPL) are measured in decibels, where a decibel is the relative value of the sound intensity or pressure compared to a reference value, based on a logarithmic scale. The IL and SPL have nearly identical values.

48. C   The heat pump system and the unitary system can have a meter on each unit, thus permitting separate billing. The multizone system uses a separate mixing box at the plant for each zone. This does not lend itself to separate metering, since mixing boxes are nearly impossible to meter and there is only one plant serving all the zones.

49. A   A step down transformer is used to transform high voltage, such as that at a power line, to a lower voltage appropriate for building use.

50. D Relative humidity is the ratio of the actual amount of water vapor in the air to the maximum amount that the air could contain at the same temperature. It is usually expressed as a percentage.

51. D Choice I is not correct. The zone of the earth that supports human life (five miles into the earth's crust and five miles into the atmosphere) is an extremely fragile ecosystem. The biosphere that has evolved over millions of years, is dramatically affected by the growth of human activity in the last 150 years.

Choice II is not correct either. While innovative technologies are improving energy efficiency of some building systems, the vast majority of the built environment is energy inefficient.

Choice III is also not correct. Toxic substances have the tendency to expand and affect large areas. For example, the air above the Great Lakes contains evidence of DDT, a toxic pesticide banned in the United States decades ago. It was discovered that DDT is captured in the jet stream bringing toxic materials from far away continents that still use toxic pesticides.

Choice IV is not correct. While recycling is helpful, it is just the beginning of the sustainable design process. The principles of sustainable design say that we need to have more building products that can be recycled and are biodegradable to create a more sustainable ecosystem.

52. D All of these consultants (wetlands engineer, energy commissioner, landscape architect, and energy modeling engineer) might be necessary for the holistic approach to sustainable design. The landscape architect should have experience with local, native plant design.

53. A I is correct. Solar shading, whether from landscaping or architectural elements, can regulate the insulation to increase winter light and reduce warm summer sunlight.

II is also correct. Urban heat island effect is the tendency of a building roof to absorb solar radiation during the day and then emit heat radiation during the evening. Roof systems with grass or light colored roofing material reduce the urban heat island effect.

III is not correct. Sustainable design encourages approaches that reduce the area allocated to parking.

IV is correct. The type, location, and size of building fenestration are a keys aspect of architectural design for sustainable projects.

54. A All elevators used by the public must comply with the ANSI handicapped standards. For example, all cars must operate automatically, which includes self-leveling devices (I) that align cars with floor landings within a 1/2 inch tolerance, as well as door reopening devices (II) that automatically function if the door becomes obstructed by an object or person. However, elevator cars do not require emergency communication systems (III), and car control buttons may be mounted up to 54 inches, not 36, above the floor (IV). Finally, elevator car floors may be covered with any stable and regular finished surface, including carpet and padding (V); however, such carpeting should have a short pile (1/2 inch maximum) and be firmly installed. The correct combination of features is found in answer A.

55. B Among the many advantages of escalators are those factors listed in correct answer B. Escalators are not, however, dependable in the event of a fire. Their mechanisms may be damaged and they may fail to operate. Escalators may employ fire-protective devices, such as a sprinkler system or rolling steel shutters that close a stairwell, but these devices are difficult to incorporate in an open area, and thus, are not commonly used. Escalators are also not very flexible architecturally, as they are custom made for each installation, and once positioned, they are almost never moved to another location.

56. B Interval is defined as the average time between successive departures of cars from the main floor or lobby of a building. Although the requirements and functions of high-rise buildings vary, the accepted interval for good service in the average building is 30 seconds.

INDEX